INTERNATIONAL ENER

CW01329372

ENERGY
MARKET
EXPERIENCE

LESSONS FROM LIBERALISED ELECTRICITY MARKETS

INTERNATIONAL ENERGY AGENCY

The International Energy Agency (IEA) is an autonomous body which was established in November 1974 within the framework of the Organisation for Economic Co-operation and Development (OECD) to implement an international energy programme.

It carries out a comprehensive programme of energy co-operation among twenty-six of the OECD's thirty member countries. The basic aims of the IEA are:

- to maintain and improve systems for coping with oil supply disruptions;
- to promote rational energy policies in a global context through co-operative relations with non-member countries, industry and international organisations;
- to operate a permanent information system on the international oil market;
- to improve the world's energy supply and demand structure by developing alternative energy sources and increasing the efficiency of energy use;
- to assist in the integration of environmental and energy policies.

The IEA member countries are: Australia, Austria, Belgium, Canada, the Czech Republic, Denmark, Finland, France, Germany, Greece, Hungary, Ireland, Italy, Japan, the Republic of Korea, Luxembourg, the Netherlands, New Zealand, Norway, Portugal, Spain, Sweden, Switzerland, Turkey, the United Kingdom, the United States. The European Commission takes part in the work of the IEA.

ORGANISATION FOR ECONOMIC CO-OPERATION AND DEVELOPMENT

The OECD is a unique forum where the governments of thirty democracies work together to address the economic, social and environmental challenges of globalisation. The OECD is also at the forefront of efforts to understand and to help governments respond to new developments and concerns, such as corporate governance, the information economy and the challenges of an ageing population. The Organisation provides a setting where governments can compare policy experiences, seek answers to common problems, identify good practice and work to co-ordinate domestic and international policies.

The OECD member countries are: Australia, Austria, Belgium, Canada, the Czech Republic, Denmark, Finland, France, Germany, Greece, Hungary, Iceland, Ireland, Italy, Japan, Korea, Luxembourg, Mexico, the Netherlands, New Zealand, Norway, Poland, Portugal, the Slovak Republic, Spain, Sweden, Switzerland, Turkey, the United Kingdom and the United States. The European Commission takes part in the work of the OECD.

FOREWORD

In ever more globalised and automated economies the role of electricity is increasingly important as a driver for economic prosperity. Reliable and affordable supply of electricity is essential for the competitiveness of global industrial product markets and a necessary ingredient in the daily workings of modern societies. At the same time, environmental impacts of energy usage are one of the most difficult global policy challenges. Reliable access to affordable electricity supply with acceptable environmental impacts is only achieved with comprehensive and carefully balanced policy actions to establish the necessary incentive-based framework. To that end, liberalisation of electricity markets is a development path and policy option that has been implemented or considered in all IEA member countries.

Through competition in liberalised markets incentives are created to drive for more efficient operation of electricity systems and more efficient investment decisions in terms of timing, sizing, siting and choice of technology. Even if liberalised markets leave critical policy challenges un-resolved, the transparency created by competition tends to improve the framework for targeted policy actions to address issues such as environmental quality and reliability.

After up to ten years' experience with liberalised electricity markets and even longer in some cases important lessons can now be drawn from some pioneering countries and regions. Theoretical principles for successful liberalisation can now be augmented with more qualified policy prescriptions based on real-world experience.

Even if some pioneering markets have operated with considerable success for a number of years, liberalisation has shown it self not to be a single event, but rather a long process that requires on-going government commitment. No markets are perfect, and they will continue to evolve and develop to match the needs of electricity systems - systems that are at the same time undergoing considerable change.

This book addresses the main principles of successful liberalisation with actual experiences and outcomes, hopefully providing decision makers within government and industry with policy prescriptions on the key issues. One point of departure is to ask whether liberalisation of electricity markets is feasible – that it, has it been possible to develop a functional market without jeopardizing reliability and other public policy priorities. Secondly, if liberalisation has worked in that it has been able to achieve this balance, has it delivered the expected outcome in terms of real economic benefit. An affirmative answer to these questions leads to the books' focus on what issues are critical and what approaches best lead to successful electricity markets, allowing the book to point to best practices.

This book is the first in the IEA series on electricity market experiences. It is published under my authority as Executive Director of the International Energy Agency.

Claude Mandil
Executive Director

ACKNOWLEDGEMENTS

The principal author of this book is Ulrik Stridbaek of the Energy Diversification Division, working under the direction of Ian Cronshaw, Head of Division, and Noé van Hulst, Director of the Office for Long-Term Co-operation and Policy Analysis.

This book has benefited greatly from suggestions and comments by Doug Cooke, from the IEA, who particularly contributed to issues of transmission investment and performance. Dong-wook Lee and Josef Sangiovanni from the IEA contributed to the annexes, in particular. Valuable comments were provided by Jolanka Fisher, from the IEA, Mats Nilsson, from the Swedish Energy Markets Inspectorate, Henning Parbo, from Energinet.dk and delegates from the IEA's Standing Group on Long-Term Co-operation. Muriel Custodio managed the production of the book, Corinne Hayworth, designed the layout and the front cover, and Marilyn Smith edited the book.

TABLE OF CONTENTS

LIST OF TABLES

LIST OF FIGURES

EXECUTIVE SUMMARY

Over the past decade, several IEA member countries embarked on a policy path of market liberalisation of the electricity supply industry. Pioneers in electricity market reform have now been operating with considerable success for a number of years, delivering substantial benefits to economies. Finding the most effective way to develop competitive electricity markets that fulfil the goals of real economic benefits has not been clear, however. Scepticism and concerns are voiced in many countries, and debate continues on several key issues. The sceptics point to the California crisis and market breakdown in 2001, only a few years after the new market was launched under much publicity and the subsequent and spectacular bankruptcy of the large energy company Enron. The widespread blackouts in North America, Italy and Scandinavia in 2003 are also sometimes used to argue that electricity market liberalisation is a failed concept.

Today, extensive expert opinion and research material identifies the root causes of some of the past failures. The California crisis can be attributed to a wide range of factors, including important regulatory failures in the set-up of the California market. Official investigations into the 2003 blackouts do not blame liberalised markets for being the root causes of those events. Learning from these blackouts, a recent IEA study on transmission system security points out that liberalisation has fundamentally changed the use of transmission systems. The management of transmission systems still needs to adapt to these changes. With that adjustment in place, liberalised markets can provide a framework that improves system security, largely through increased co-operation amongst jurisdictions.

While the public has focused on the remarkable failures of the past decade, several electricity markets have been operating successfully and have developed into robust markets during the same period. In all IEA member countries, the liberalisation process has progressed at varying speeds. Despite the fact that no straightforward path to success has emerged, there is a general lesson to be learned: Electricity market liberalisation is not an event. It is a long process that requires strong and sustained political commitment, extensive and detailed preparation, and continuous development to allow for necessary improvements while sustaining on-going investment. It is, in fact, a process that has not yet been completed anywhere in the world - nor will it be in the foreseeable future.

This book focuses on the experiences and lessons learned in several pioneering markets that have now operated with considerable success for a decade or more. While discussing prescriptions for successful electricity market liberalisation and drawing on real-world experiences, three questions are pursued: *Has electricity market liberalisation been practicable? Has it delivered real economic benefits? If so, what are the key issues for success?* The key messages, conclusions and recommendations following from the findings are presented in this Executive Summary.

Electricity Market Liberalisation Delivers Long-term Benefits

Traditionally, electricity sectors developed and operated within strictly regulated frameworks in which vertically integrated utilities have handled most or all activities – from generation to transport to distribution. Moreover, it has been a centrally planned activity, wherein needs are assessed and fulfilled by electricity system planners and all associated costs are passed on to consumers.

But traditional, vertically integrated utilities tend to create substantial overcapacity, a fact that became more obvious when electricity demand growth slowed during the 1980s and 1990s in many IEA member countries. In addition to reducing this overcapacity, liberalisation has also been shown to provide large potential gains from improved efficiency in the operation of generation plants, networks and distribution services.

Monitoring of electricity rates paid by different customer classes is one basic way to assess the performance of liberalised electricity markets. Indeed, many countries promised falling prices prior to launching liberalisation processes. In markets that have liberalised successfully, there is a clear trend of falling electricity prices for industrial consumers, in both nominal and real terms. The trend is less clear, and certainly slower, for household consumers. However, prices paid by consumers do not necessarily reflect the costs of producing and transporting electricity. Some consumer groups often subsidise other consumer groups. Different parts of the value chain – from the recovery of fuels to generation and transport of electricity – are also often subsidised in one way or another, or are not fully cost reflective for other reasons. Electricity rates and taxes are often related in non-transparent ways. Changes in fuel costs and environmental regulation, affect final costs of supplying electricity and seems to be important drivers for recent increases in electricity tariffs in many particularly European IEA member countries, but are not directly related to the effects of electricity market liberalisation. In addition, investment decisions made within a vertically integrated industry influence electricity costs for a long time, hence the effects of past investment decisions will be reflected in retail prices for several years to come. All in all, these factors make electricity retail prices paid by end-users complex to interpret. In reality, retail prices are poor indicators of whether performance development is positive in the electricity industry.

Examining performance in various specific segments of the value change paints a clearer picture. For example, many countries mothballed unused generation assets and data show that they now use existing plants more efficiently. At the same time, fundamental changes in the use of transmission assets has created more dynamic and enhanced usage, often resulting from increased trade across jurisdictions. Other indicators show marked increases in labour productivity.

A recent study by the Organisation for Economic Co-operation and Development (OECD) explores the benefits of liberalising product markets and reducing barriers to international trade and investment across several regulated sectors. It singles out electricity as one of the sectors with the greatest potential for improvement. The results of the analysis assess the total annual benefits across all sectors to be 1% to 3% of GDP in the United States and 2% to 3.5% of GDP in the European Union. The study assesses only the static benefits from increased trade and better allocation of resources, but recognises that dynamic gains from increased innovation may be even greater.

According to traditional performance indicators, electricity market liberalisation is already delivering significant benefits. But it is perhaps even more important to consider how a liberalised electricity sector is better equipped to meet challenges and exploit opportunities of a future more diverse and flexible portfolio of technologies. Combined Cycle Gas Turbines is a preferred generation technology option in many circumstances which is re-enforced in the new market framework where focus on operational and financial flexibility has increased. Uncertainty, which creates financial risk, is seen in a new light; today's market players show a preference for less capital-intensive and smaller units. Technological developments are also creating opportunities for greater consumer involvement in decisions that determine the system. Another driver for change in the technology mix comes from government support of a wide range of technologies - many for their environmental merits and some of which are already being deployed at high rates. Nuclear power is re-surfacing as a technology that warrants serious consideration in some countries.

All in all, a more diverse electricity system is beginning to emerge. Correspondingly, management of this system must accommodate many players, ranging from very small generation units and demand resources to very large nuclear power plants. Thanks primarily to transparent price signals, liberalised markets are creating a more level playing field and allowing for the necessary co-ordination of many diverse market players. As liberalised markets begin to mature, it becomes more obvious that a centrally planned and vertically integrated approach is less appropriate for such a diverse system and is, in fact, likely to be a barrier to the innovation required to meet the future need.

For the moment, it is critical to avoid being overly short sighted. Liberalisation is expected to bring large economic benefits for consumers and societies in the long term and evidence so far indicates that markets can deliver these benefits. But in the short term certain groups may not realise immediate benefits or may even experience losses. Vertically integrated utilities are likely to feel threatened by the requirement to unbundle. Consumer groups that previously benefited from subsidised electricity tariffs (at the expense of other consumers) may perceive liberalisation as a loss as cross-subsidies are unwound. Certain segments of the utilities' workforce will feel threatened when open competition demands higher efficiency and increased labour productivity. Without question, one of the most critical policy challenges facing decision makers is the management of social and equity issues in distributing the benefits of electricity market liberalisation.

Key Message

Electricity market liberalisation has delivered considerable economic benefits.

Under pressure from competition, assets in the electricity sector are used more efficiently, thereby bringing real, long-term benefits to consumers. Liberalisation to introduce competition is, however, a long process rather than an event: it requires on-going government commitment to resolve challenges when vested interests and cross-subsidies are unwound.

Government has a Critical but Fundamentally Changed Role

Regardless of the approach to liberalisation, the process requires strong government involvement. In fact, the level of on-going political commitment invested significantly influences the outcome. In the absence of clear signs of commitment, regulatory uncertainty may well become self-fulfilling and undermine a positive outcome. From time to time, all electricity systems will experience a crisis. Such crises have become important tests of the robustness of liberalised electricity markets and, perhaps even more importantly, of the robustness of the political framework backing the liberalisation process. At difficult junctures in market development, signals of strong political commitment - often expressed by *not* intervening - can lead to the necessary market responses.

Effective markets are fuelled by competition. Thus, one of government's most decisive roles is to establish a framework that allows for the development of effective competition. Liberalisation creates benefits by introducing incentives for higher efficiency and more innovation - with competition as the driver.

The first step required to introduce competition is to break down the monopolies that exist in traditional vertically integrated utilities. It is necessary to separate network activities from all other activities, either through legal unbundling of the network entities or, more effectively, through true ownership unbundling. The key is to introduce competition in as many parts of the value chain as possible - from generation to consumption. Remaining natural monopolies (*e.g.* networks and system operation) should be subject to continued and improved economic regulation.

Unbundling effectively breaks up the centralised decision-making process found in vertically integrated utilities, replacing it with a decentralised process where market players make decisions within markets. This can only work smoothly when markets are 'effective', but effective markets do not develop automatically. Creating a level playing field and developing effective, competitive marketplaces requires establishing detailed market rules, design and regulation. Within the on-going liberalisation processes, the level of government involvement through detailed legislation and rule-making has varied. But it is evident that governments are critical to establishing a framework with the necessary incentives. At the same time, independent regulators are one of the critical bodies within this framework; their role in overseeing compliance with legislation and ensuring fair and efficient economic regulation of networks is fundamental to successful market development.

Real-time system operation is an aspect of the electricity sector that is maintained as a natural monopoly and, thus, should be unbundled from other competitive segments of the value chain. Market rules, design and regulation aim to direct all actions transparently, but many subtleties remain in secure, day-to-day system operation. It is inevitable that system operators will preserve certain discretionary powers, regardless of careful efforts to regulate grid access. Their independence is particularly critical to the creation and further development of well-functioning and robust markets.

In the new decentralised industry structure, transparency is a prerequisite for developing competitive liberalised electricity markets. Competitive market players do not automatically (or voluntarily) collect and publish fundamental market data and statistics. Therefore, it is important to redefine responsibility for this necessary task in liberalised markets. Increased transparency is a proven, strong instrument to ensure continuous development towards more effective markets. In fact, transparency adds to the benefits of liberalisation in its own right, by improving the decision-making framework for all actors - policy makers, industry and consumers alike.

But a formal framework that allows for competition and creates a level playing field is not enough. Competition will flourish only if multiple players compete in the market. Governments and regulators have managed to enhance competition through various means, but a high level of market concentration remains a serious concern in several markets. Effective markets and transparency have been vital to easing access for new-comers. In addition, extending markets across countries and regions helps enable the "import of competition"; this is particularly important in smaller jurisdictions in which the need for consolidation limits the number of market players that can operate efficiently. To date, achievements are more limited in *ex post* regulation of competition. It is illegal to exercise market power, but it often remains difficult to prove such behaviour. In some cases, dealing with market power abuse has been further complicated when the largest companies are regarded as national champions or provide substantial revenue streams to their public owners.

Some claim that market failures are inherent across the value chain in electricity markets requiring government intervention. But, upon closer scrutiny, many alleged failures turn out rather to be the result of regulatory failures. In the event of real market failures - as might arise from concerns about reliability of supply and the environmental impacts of electricity production - governments may be called upon to intervene in more active ways.

Analysing reliability of supply in unbundled electricity sector has also called for an "unbundling" of the concept of reliability of supply into its relevant parts of the value chain. Concern has been voiced about secure supply of fuel for power generation, adequacy of investment in generation and network assets, and the security of real-time system operation.

Considerations concerning market failures involving the secure supply of energy or fuel for electricity generation are an issue with geo-political dimensions. Increased use of natural gas as a fuel for power generation raises the issue of IEA member countries' dependence on natural gas supplies from non-member countries - and stresses the importance of developing, in parallel, a competitive market for natural gas. Thus far, the levels of security of fuel supply ensured by commercial market players have not undermined the performance liberalised electricity markets.

Also, so far, the efficiency of liberalised markets has sufficiently incentivised adequate generation capacity. Despite efforts to integrate some network investment decisions in the competitive market framework, networks remain, more or less, regulated monopolies. Adequate network assets must be incentivised through a framework of economic regulation.

Finally, when it comes to secure real-time system operation, markets so far have failed to provide a complete framework of incentives without jeopardising system security. Government intervention is necessary and this has been carried out (rather effectively) through the establishment of truly independent system operators. A recent IEA publication explores the main requirements and challenges for effective regulation that creates the necessary levels of transmission system security.[1]

The environmental effects of electricity generation are not addressed by normal incentives in competitive markets. Environmental benefits are classical public goods and liberalised electricity markets will not adequately account for their value - or for the cost of their potential loss. Policy intervention is needed to ensure they are properly taken into account. Policies motivated by environmental and climate change concerns are already having serious impacts on liberalised electricity markets, as was intended.

Many environmental policies are, however, potentially distortive beyond the initial intent, particularly when looking across internal markets within the context of international competition. Direct financial support for particular technologies, or non-transparent barriers that block development of others, can lead to inefficiencies and distort competition. This adds uncertainty to the investment decision process and ultimately poses a threat to the system. In several liberalised electricity markets, the preferred option to address this issue is implementation of cap-and-trade policies. This approach transforms the political goal into an obligation imposed upon market players. Market players are then left to fulfil the obligations in ways that they find optimal, including trading the obligations amongst themselves and finding alternative technology solutions.

---

1. *IEA (2005)*, Learning from the Blackouts: Transmission System Security in Competitive Electricity Markets.

For example, carbon dioxide (CO_2) emission trading was introduced in Europe to address climate change. Creating a market for CO_2 emission permits creates a price for CO_2 emissions, thereby making it possible to internalise the environmental cost within the total cost of producing electricity. Following similar principles, support for renewable energy has changed from direct financial support for specific technologies to market-based certificate systems backed by obligations. Fulfilling climate change policies through cap-and-trade instruments seems to be the least distortive option. In addition, this approach adds transparency that enhances the quality of decisions made by policy makers and industry.

Key Message

Establishing truly independent and committed regulators and system operators precedes a competitive framework.

Liberalisation still requires strong and committed government involvement, but in fundamentally changed roles. Vertically integrated utilities must be unbundled. Governments, independent regulators and independent system operators must collaborate to establish rules, market design and regulation that create a competitive market place and support its further development. Continued government commitment and signs of support is crucial, particularly at the difficult junctures every market will, inevitably, face.

Price signals are the glue

In the process of unbundling utilities to introduce competition, vertical integration has been replaced with markets comprising multiple players. In this new framework, price signals direct decisions in the marketplace. Efficient decisions depend on correct signals, *i.e.* price signals that reflect the real costs, benefits and values of producing, transporting and consuming electricity.

Electricity has a value to the consumer only if it is supplied at the right place, at the right time, in the right volume and at an acceptable quality. The locational aspect of electricity pricing is the most controversial and complex issue in efficient pricing. Principles that establish a price for each node in a system are the ideal reference because they value electricity based on where it is generated and delivered, and some markets come close to achieving this.

However, there are important trade-offs to consider when choosing pricing principles that could justify a less fine-tuned, zone-based system, where a price is established for several nodes that are rarely congested. Even though there are important trade-offs, the main controversy often relates more to social equity and distribution rather than specific pros and cons of market functioning and system operation. Nodal pricing evolved as a necessity in highly meshed networks where transmission lines are criss-crossing the electricity system (*e.g.* North America); zonal pricing is accepted as a good approximation in more radial networks, where the structure of congestion is less complex (*e.g.* Australia). Higher transaction costs and the greater complexity of nodal pricing are often used to argue for pricing principles that are less reflective of location. In reality, evidence shows that obvious congestion points have often not been priced appropriately. The highly meshed network in continental Europe is currently developing into a zonal market, often with countries constituting entire zones, thereby potentially bluring price signals and inhibiting efficiency.

Open trade across jurisdictions is one of the classical merits of liberalised and competitive markets. It enables exploitation of comparative advantages - at mutual economic benefit for all regions involved. Electricity generation and transport include many factors related to resource endowments, geographical characteristics and regional skills. They are both very capital-intensive businesses that can realise significant gains by optimising asset use across as large an area as possible. This is particularly true for smaller jurisdictions. But trade across jurisdictions relies on co-operation amongst system operators. Therefore, independence and appropriate incentives of system operators are critical in the development of cross-border trade.

Designing an appropriate market model is a matter of creating trading arrangements that fit to the specific circumstances of each electricity system while also addressing broader key issues of unbundling, third-party access, cost-reflective pricing and transaction costs. There is not one single winning market model - no one-size-fits-all – in each situation important trade-offs must be made. However, there is one common feature of all successful markets: some sort of formal price quotation, conceived through formal market design. Cost-reflective pricing principles are an aspect that must be fostered through regulation and active market design and will not develop without well-designed market rules.

Electricity consumption and supply are inherently volatile. But the volatility is an inseparable characteristic of the service and is not related to the organisation of the sector. Liberalised electricity markets create a more transparent framework, allowing for cost-reflective pricing reflecting this volatility. In some instances, government interventions to suppress volatility and cap prices below what can be justified by economic reasons have blurred price signals and slowed market responses.

Price volatility creates risks for market players, including generators and consumers. Risks are the result of uncertainty, and there is considerable uncertainty connected with many of the fundamental factors that determine electricity generation, transport and consumption. In the previous model of a vertically integrated and regulated sector, all costs - and, therefore, all risks - could be passed on to consumers. Liberalised markets make risks more transparent and, more importantly, re-allocate these risks to the decision makers themselves.

In liberalised electricity markets, business risks can be effectively managed through contracts. Generators, retail suppliers and consumers can agree on prices, volumes, times and other conditions that create the desired certainty within the framework of the contract. In fact, liquid and effective markets for financial contracts improve competition by enabling sophisticated risk management. This, in turn, eases market access for new and smaller market players and contributes to ensuring that market power is not exercised. Most markets provide a framework for a liquid market in the day-ahead and real-time segments through market rules and design. In some markets, relatively liquid and effective financial markets for longer-duration contracts are developing, but the evolution of these markets remains a major concern for the creation of robust electricity markets.

Key Message

A framework for efficient and cost-reflective prices is established through regulation and market design.

Prices that reflect the inherent volatility of electricity are critical in creating a framework for efficient decisions regarding operation and investment in liberalised markets. Electricity market design, as conceived and implemented by governments, independent regulators and independent system operators, can create prices that reflect real costs through mechanisms that incorporate time, volume and location. Blurring price signals with price caps or insufficient locational signals slows market responses in short-term operation and longer-term investment. It is therefore potentially risky for reliability, particularly in the longer term.

Empowering the Consumer

Vertically integrated utilities naturally focus on the supply side of the electricity sector, concentrating on the two pillars of electricity generation and transport. Until now, consumers paid the bill, and no infrastructure was in place to involve them in decision-making processes. Liberalised electricity markets introduce a third pillar that allows consumers to become active participants. Effective markets allow consumers to exercise their right to shift supplier, thereby enhancing competition for better services and increased innovation. Perhaps more importantly, consumer response to prices adds real resources to the system, potentially saving expensive generation or transmission investment and improving reliability. Finally, improved transparency from cost-reflective prices provides clearer incentives for more efficient energy use. This new third pillar is a product of the recent liberalisation process and, unsurprisingly, it has developed more slowly than the other two pillars. While the framework for consumer participation now exists, many of the detailed structures needed to facilitate ease of participation must still be further developed.

A first building block to empower the consumer to participate is to create the necessary competitive pressure. Such pressure creates the incentives needed for retail companies to bring the opportunities of a competitive wholesale market to the doorsteps of consumers.

Retail competition is based on the same principles as wholesale competition, with regulated access to the grid being the key factor. Unbundling of competitive retail activities from network activities is an important aspect of the process, but in most cases this phase of liberalisation has been less comprehensive than in transmission and system operation. Regulated access is provided by constructing systems and formal rules for consumer switching, but many markets still have small, but possibly decisive, barriers to switching - or still offer advantages to incumbent semi-integrated retail and network businesses. In all competitive markets, larger industrial consumers have switched in great numbers. The experience for smaller commercial and residential consumers is more varied, ranging from high switching rates in some markets to disappointingly low rates in others. In jurisdictions with liquid financial markets, more sophisticated retail products have been developed to better serve the needs of consumers who want to take an active role in managing risks. However, overall product innovation and development has been slow and sporadic.

Anticipating that many small commercial and residential consumers will not switch retailers, some states and countries opted to introduce a regulated default tariff or contract to protect consumers. While the justification for such regulated contracts is clear in a political context, they also often create distortions that may serve consumers poorly in the end. Establishing competitive retail markets that provide easy access to switching between competing retailers remains a challenge.

Another effect of the somewhat slow development of competitive and innovative retail markets and the still often supply focused market design is the failure to bring wholesale market prices to the doorsteps of consumers. So far, there has been only limited opportunity for consumers to create benefits by shifting load as a price response. Considering that electricity is consumed by millions of different consumers for millions of different purposes, consumers are undoubtedly, in principle, willing to shift demand by varying degrees as a response to different prices. Demand is price-elastic: the challenge is to lower transaction costs sufficiently to justify participation for consumers who stand to realise the largest potential benefits. At present, a large barrier is in the hands of retail companies that must bring wholesale prices to the consumers' doorstep. That being said, there must also be something to respond to: consumers cannot be expected to respond before prices rise sufficiently to off-set transaction costs. The largest consumers, who already have remotely read interval meters, are likely to be the first to see the benefits of shifting demand in response to price. Finding a way to take the wholesale price to the doorsteps of smaller commercial and residential consumers is, however, fraught with a technical and economic barrier given the absence of necessary metering equipment.

Lack of demand participation remains one of the most serious challenges in liberalising electricity markets. The barriers are numerous. Creating easy and effective systems to manage retail switching is challenging. For small residential consumers, the infrastructure to enable switching is relatively costly compared to the potential benefits. In addition, it has been difficult to remove all distortions from semi-integrated networks and retail companies. Where governments show a willingness to intervene through price caps and other means, this also serves as a barrier to demand participation. Finally, lack of liquid financial markets makes it difficult to create the necessary innovative products. However, early evidence shows that consumers do switch suppliers and do respond to price when the conditions are sufficiently good. In fact, remarkably little demand response to price is necessary to significantly improve the performance of electricity markets, enhance system security and substantially reduce volatility and electricity prices for all consumers.

Key Message

Removing barriers to retail-switching and to active demand participation empowers consumers.

Liberalised electricity markets provide a framework that can empower the consumer to exercise freedom of choice, to realise benefits from active response to prices and to consume energy more efficiently. Effective retail competition is a prerequisite to realising these benefits. A formal system must be in place to enable quick and easy retailer switching. Demand resources must also have easy access to wholesale, balancing and reserve markets. Opportunities to install metering equipment and infrastructure that serve the needs of different consumer groups should be considered, with due attention paid to cost-effectiveness.

Efficient Incentives for Investment are Critical

A substantial share of the electricity consumer's bill goes toward financing generation and network assets. The opportunity to improve investment decisions is a significant potential benefit of market liberalisation. The ability of electricity markets to provide sufficient incentives for timely and efficient investment in generation plants continues to be one of the most debated aspects of market design. Many investment projects require long lead times and have an economic lifetime of several decades. The transitional phase of market development is characterised by uncertainty that may undermine the investment climate - and ultimately the successful transition to a competitive market. Investments in power generation are one of the big tests for the development of robust markets.

Liberalised markets create a new investment paradigm in which decisions are taken under competitive pressure. When risks are shifted from consumers to decision makers, capital-intensive technologies with long construction times are viewed with greater scepticism - even if marginal costs are low. In the new competitive environment where risks are transparent, market players prefer technologies with short leadtimes that can be built in small incremental steps. Competition also pushes investment decisions to the last minute, which saves resources but can also put policy makers under pressure to intervene in a transitional phase (*i.e.* before the process has proven to be robust).

In situations where the supply and demand balance is tight, demand response to price can constitute the necessary buffer of last resources and add much-needed flexibility. To date, a certain level of active demand participation has been critical in re-affirming the robustness of markets; conversely, lack of demand participation has laid the groundwork for very high price spikes needed to trigger investment. When governments have refrained from intervention and let prices reflect real costs, markets have come through – they have not failed to provide incentives for a response through investment in new generation capacity. In this context, so-called "energy only" (or, more correctly, "one price only") markets, in which the wholesale electricity price provides remuneration for both variable and fixed costs, have performed well.

Some markets have not shown enough confidence to rely on the delicate balance inherent in this new investment paradigm. These markets assume that consumers are not willing to participate and thus find that protective price caps are necessary as a consequence. However, with the barrier of a price cap, extra incentives must be added to prompt timely and adequate investment. These extra capacity measures have been implemented in various forms and have incentivised new investment. But they have also been prone to market manipulation. Another drawback is that capacity measures force decisions regarding the overall need for new generation capacity back into a centralised decision-making process.

Investments in networks are, by and large, still being made within regulated frameworks. Distribution networks are still regulated, even though liberalisation reinforced the focus on cost-cutting in the whole value chain, including networks. The introduction of new economic regulation models prompted cost cuts but, in many countries, the focus is now shifting to quality aspects. Low investment levels and quality problems have directed the focus towards the necessary regulated rates of return to award investment and other incentives tied to quality.

The development of regulated transmission networks is similar to that of distribution networks. Locational pricing in wholesale markets has, however, added crucial analytical input to enable more qualified investment decisions in transmission networks. In effect, transmission networks are an alternative to generation plants. Thus, in principle, transmission investments could rely entirely on locational pricing and commercial, competitive enterprise. In reality, the business model for merchant lines has proven to be fragile. Very few merchant lines are currently financed by purely commercial means, but locational pricing has still added substantial transparency to the process of making investment decisions in transmission. For example, several markets are developing information systems that enable a more co-ordinated interaction between decisions on regulated transmission investments and decisions on investments in generation plants.

It is important to design markets and create regulatory frameworks that provide sufficient remuneration and incentives for efficient investment. But none of this makes any difference if investors cannot get permission to build. The absence of transparent and smooth approval procedures - whether to use a particular technology or to site a new generation plant or network at a particular location - continues to be a serious barrier to investment in most markets. This is not related to the liberalisation of electricity markets; rather, cost-reflective locational prices make the consequences of related environmental policies and the so-called "not in my back yard" (NIMBY) syndrome more transparent.

Key Message

Minimising regulatory uncertainty is key to creating a framework for timely and adequate investment.

Some amount of regulatory uncertainty is a fact in a changing world, but governments can take a number of measures to credibly minimise this uncertainty for investors. Investors in new generation capacity, particularly small market players, need access to market information, a well-established marketplace and regulated access to this marketplace. An investor will also require a transparent and clear procedure for applying for new investments, in terms of both siting and choice of technology. Any signal of government willingness to intervene in the market, including possible future support and possible price capping, will add uncertainty and deter investment. Finally, regulated network investment requires adequate regulated rates of return to maintain an acceptable quality of service in the long run.

INTRODUCTION

For more than a decade, market liberalisation has been a central policy path in the electricity supply industry in several IEA member countries. The movement to privatise and introduce competition in traditionally state-controlled and regulated sectors came to the electricity sector at the end of the 1980s and early 1990s, with several IEA member countries pioneering the way with considerable success. Since then, virtually all IEA countries have initiated liberalisation processes. However, the path towards competitive electricity markets - the goal of which is to deliver real economic benefits - has not been clear. In many countries, policy makers, practitioners and academic experts continue to debate several key issues while other stakeholders voice scepticism and concern.

Sceptics point to the crisis and market breakdown that hit California in 2001, only a few years after the new market was launched amidst much publicity. The crisis led to some very visible consequences for consumers, who were disconnected from the grid in rotating blackouts, and for electric utilities, which suffered major financial distress and, in some cases, were forced to file for bankruptcy. The collapse of the Ontario market shortly after its launch in 2001 is also singled out as an example of failed market reform.

Another spectacular case is, of course, the bankruptcy of Enron. By being at the forefront of market analysis, trading and risk management, Enron mastered the emerging global electricity markets and successfully competed to become one of the largest energy companies in the world. Its foundation was, however, undermined by fraud, and alleged ruthless abuse of market power by the largest generators in California, including Enron, may have contributed to the collapse of that market.

In the eyes of the general public, the large blackouts that hit North America, Italy and Scandinavia in 2003 are perhaps also regarded as examples of the failure of electricity market liberalisation.

But extensive expert opinion and diverse research material now point to the root causes of those events, suggesting that the Californian crisis should be attributed to a wide range of factors - including both unfortunate coincidences and other more fundamental problems. For example, the "not in my back yard" (NIMBY) syndrome obstructed new construction work that was needed to meet demand. Much of the literature shows broad agreement on one point: the framework of the California market had important regulatory failures - on some very critical accounts. In essence, when policy makers negotiated the final market design, the regulatory framework failed to properly account for some key principles linked to the nature of competition and economic incentives. The criminal acts for which Enron and other utilities now face charges only accentuated an already existing crisis.

The alleged connection between the large blackouts in 2003 and electricity market liberalisation has also been explored extensively. Official investigations of the blackouts do not blame market liberalisation. The IEA points out in a new study that liberalisation has fundamentally changed the usage of transmission systems, and that the management of transmission systems had not properly adjusted to the new regulatory and market framework in the electricity sector. The study argues that, with appropriate adjustments in place, liberalised markets can create a framework that actually improves system security through increased co-operation amongst jurisdictions[2].

While a great deal of public attention has focused on these remarkable failures, several liberalised electricity markets operated with considerable success. In other IEA member countries, the failures put a halt to liberalisation processes and provoked second thoughts. But overall, liberalisation continues at varying speeds around the globe.

This book examines the experiences of liberalised markets that have been operational for a number of years. None of these markets are perfect and all will continue to develop and evolve. But current results are starting to indicate the achievement of a robust framework. The IEA published a number of books on regulatory reform and liberalisation in the electricity sector during the liberalisation period, the latest being *Power Generation Investment in Liberalised Electricity Markets*[3] and *Security of Supply in Electricity Markets*[4]. Until now, these reports focused primarily on analysing and describing key principles for successful electricity market liberalisation. The time is now right to change the focus and examine the knowledge and experience gained in solving key issues of electricity market liberalisation, as well as to tease out some of the lessons learned and some of the real-world market outcomes.

The first serious attempt to form a liberalised electricity market was launched by Chile in 1982. Today, markets in several IEA member countries have operated relatively successfully for a number of years. England and Wales pioneered the way with the first market launch in 1990. The Nordic market started in Norway in 1991, with other Nordic countries (Sweden, Finland and Denmark) joining during the second half of the 1990s. In Australia, markets in Victoria and New South Wales began operating in 1994; the Australian

2. *IEA, 2005c.*
3. *IEA, 2003a.*
4. *IEA, 2002b.*

National Electricity Market (NEM) followed in 1998. New Zealand liberalised around the same time, officially launching in 1996. In North America, various markets in the Northeast of the United States started operating in the late 1990s: PJM Interconnection in Pennsylvania, New Jersey, Maryland (PJM); the New England market; and the New York market. The California market opened in 1998. Texas in the US and Alberta in Canada opened their markets slightly later, in 2001. Additional markets are now opening across the European Union. Spain has operated a formal market based on regulated third-party access since 1998, as has the Netherlands. Many European countries report significant and growing trade over the past several years, with Germany taking centre stage.

This book focuses on experiences in selected markets that were chosen for two reasons: a) being continually operational for the longest period of time and b) being representative of the main approaches to market organisation and design. The primary examples that meet these criteria are the markets in England & Wales, the Nordic countries, Australia and PJM. Additional lessons are also presented from other markets and overall conclusions are backed by the full range of experiences. Brief descriptions of the focus markets can be found in the Annexes of this book. However, the intent is not to describe certain countries, regions or market models, but rather to highlight different ways of addressing key issues.

Although this book covers key issues for successful electricity market liberalisation, these issues are not presented in any order of priority. All items are crucial but there is a certain chronology not to be interpreted as emphasis. First, it is important to explore the main rationale for embarking upon a liberalisation process by assessing potential and experienced benefits. The second challenge relates to creating competition, the real fuel for successful liberalisation, and to understanding the roles of various stakeholders, including government. Without question, liberalisation is a fundamental change for the sector, particularly in terms of replacing the vertically integrated utilities with unbundled markets players and shifting decisions from a central body towards individual market players whose activities must be directed by transparent and cost-reflective price signals. Inherent in this transformation is the need to manage risk within this new competitive framework, which also puts the consumer in a new position. The next sections address issues associated with investment on two levels in terms of new power generation capacity (most markets face an initial phase of overcapacity) and the need for investments in networks, which seems to come sharply into focus after an initial phase in which the primary objective is to cut costs. And, finally, the book discusses the limits of liberalised electricity markets, particularly in the context of the need to ensure secure system operation and to account for environmental externalities.

ELECTRICITY MARKET LIBERALISATION HAS DELIVERED LONG-TERM BENEFITS

Traditionally, electricity sectors have developed and operated within strictly regulated frameworks in which most or all activities - from generation to transport to distribution - are handled within vertically integrated utilities. Often, there is great government involvement, both in ownership and in determining the framework. The electricity sector was typically a centrally planned activity within which needs were assessed and fulfilled by electricity system planners and all the costs were passed on to consumers.

In the late 1980s and early 1990s, a growing trend to liberalise traditionally public or regulated activities reached the electricity sector. In many countries, privatising state-owned companies is now an integral part of economic development. Within the electricity sector, the timing of the trend was motivated by inefficiencies in the sector itself and, to a certain extent, also by changing perceptions of the roles of public ownership, incentives, competition and markets. Technical development of generation technologies and - in some jurisdictions - funding problems accentuated the inefficiencies and the need for reform. Subsequent technical development enabled the management of increasingly complex systems; systems that now contain a range of the mechanisms required to support effective liberalisation of the electricity sector. In short, several factors drove the trend: poor performance of the existing systems, the growth of a "market ideology" and the declining costs of changing the system. The interaction and tension between these motives continues to drive much of the debate about benefits and costs of liberalisation in many countries around the world.

Inefficiencies in the traditional, vertically integrated utilities differed markedly from country to country, but substantial overcapacity in generation seemed to be a common characteristic. This was even more evident when electricity demand growth slowed in many IEA member countries during the 1980s and 1990s. At the time, all costs - and therefore all risks - were borne by the electricity consumer, but the utilities themselves made decisions regarding investment. Hence, incentives to balance risks, costs and benefits were effectively separated from the decision makers and the decision-making process. This was particularly important when making investment decisions in terms of choice of technology, timing, capacity and location. The costs resulting from such investment decisions are, by far, the greatest share of the total costs of electricity supply. Thus, the potential benefits from improving the decision-making process for investments are also large.

Improving decision-making processes for investment was important but improving actual operation also provided a rationale for liberalisation, specifically in terms of asset utilisation, fuel efficiency and labour productivity.

Lack of involvement on the demand side proved to be another significant source of inefficiencies in the vertically integrated and planned sector. A constant dilemma and almost inherent characteristic of the centrally planned approach is a disproportionately high focus on supply. Customers are often thought of as "connection points" or "outlets". There is no effective infrastructure in place to directly involve consumers in critical decisions that determine system design. Overcapacity and very limited product development are natural consequences of this environment. Some vertically integrated utilities developed "demand side management" (DSM) programmes, but such efforts did not offer the same dynamic interaction that is inherent in real markets. Freedom of choice of supplier adds value in its own right but, more importantly, it is also one of the main drivers for innovation and other dynamic improvements.

Finally, a focus on supply within a specific jurisdiction and guaranteed pass-through of costs also remove clear incentives for effective and dynamic trade and co-operation across national and regional borders. For smaller jurisdictions, increased co-operation in a more integrated market provides strong potential for efficiency improvements; this is one of the most important sources of benefits from liberalisation.

Indicators of Success

A natural place to begin assessing the success of electricity market liberalisation is to look at the prices actually paid by electricity consumers, including both industrial and residential customers (Figure 1).

In all of the countries included in Figure 1, there is a clear trend showing that, after market liberalisation, electricity prices fell for industrial consumers. In the United States, this downward trend began in the early 1980s, coinciding with the emergence of independent power producers (IPPs) as a first step to introduce market competition. Development of US prices as an indicator of the success of electricity market liberalisation is blurred by the fact that not all states have liberalised. Pennsylvania is also included in Figure 1 to illustrate the development in one of the PJM states. Prices in Pennsylvania are above but converging to the US average. For household consumers the trend is less straightforward. There is a clear falling trend in the United Kingdom. In Australia, prices seem relatively stable after 1998, however, households are

Figure 1

Electricity prices paid by end-consumers: Fixed 2004 prices, excluding taxes

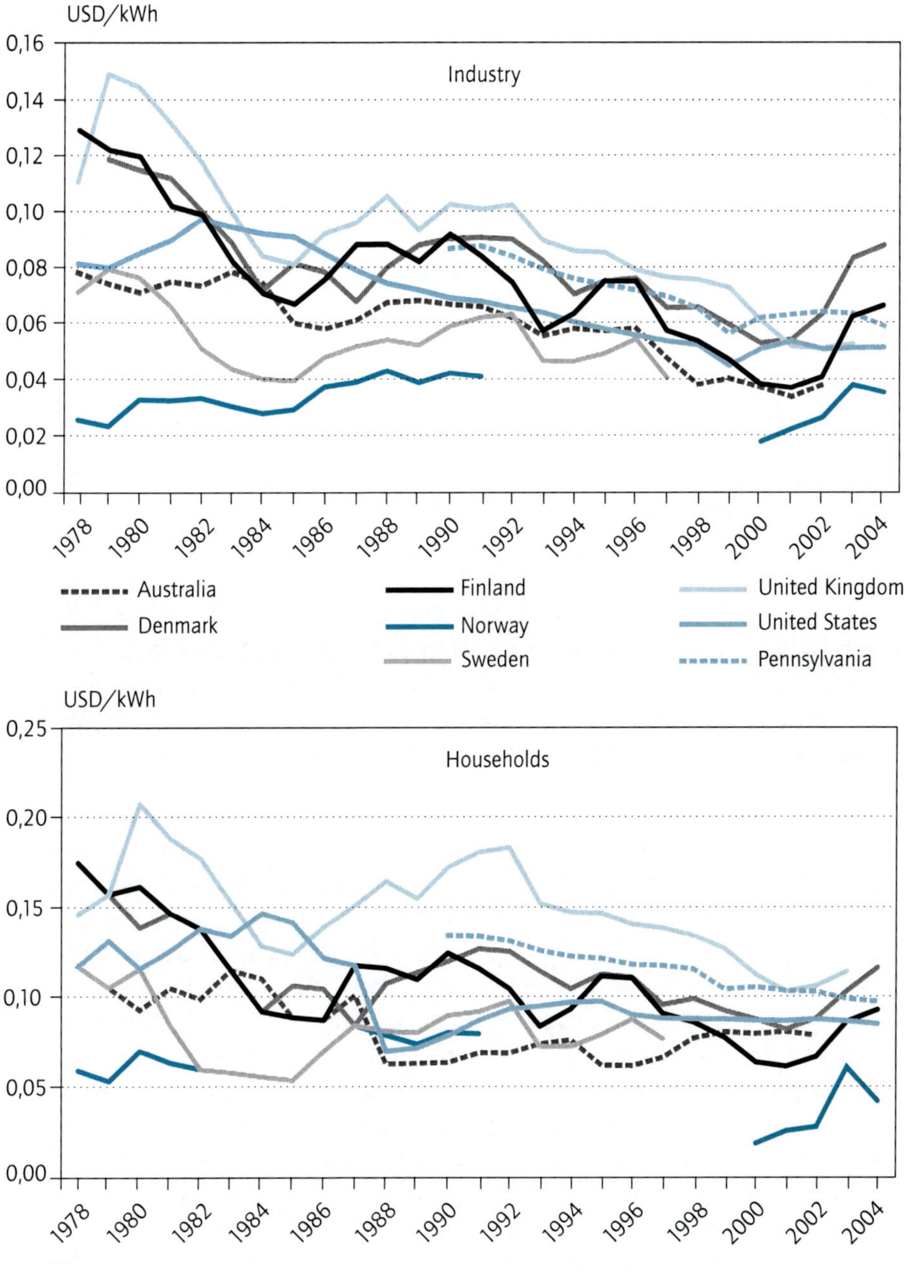

Source: IEA

only contestable (*i.e.* free to choose a supplier) in a few Australian states and this contestability was put in place only in 2001. It is interesting to note that while there is a large convergence of wholesale prices amongst the states and territories within Australia's National Electricity Market (NEM), such a trend is less evident for retail prices. In all Nordic countries, prices surged from 2002 to 2003 following a huge increase in wholesale prices as a consequence of a severe drought in the Nordic region.

Retail prices are currently increasing in many, particularly European, IEA member countries. Increasing fuel costs and costs for CO^2 emission permits in Europe seems to be important drivers but it also puts emphasis on the importance of a truly competitive framework during such periods with marked changes in market fundamentals.

Comparing prices paid by electricity consumers before and after liberalisation may seem to be a very straight-forward indicator and, indeed, many countries promised that liberalisation would deliver decreasing prices. However, prices paid by consumers are not a simple reflection of the costs of producing and transporting electricity products. Traditionally, some consumer groups subsidised others. In addition, different parts in the value chain - from fuel recovery to generation and transport - are often subsidised in one way or another, or are not fully cost-reflective for other reasons. For example, in many countries the domestic coal industry has a long history of subsidisation. Substantial government support is also invested in development of various generation technologies (*e.g.* the nuclear energy industry). In countries with government-owned utilities, taxpayers have borne some of the risks and, therefore, some of the costs.

Transmission and distribution networks contribute important components of total costs, but these result from developments in economic regulation rather than competition in liberalised markets. Another significant cost component comes from the fuel costs, such as coal and gas. Again, developments in fuel costs are not directly related to the performance of competition in liberalised electricity markets, even if currently increasing fuel costs stresses the importance of well-functioning and competitive markets for fuels. Perhaps most importantly, investment decisions made within a vertically integrated industry will influence costs for many years. Prices paid today are partly influenced by investment decisions made during the last several decades. If the electricity sector develops in new directions, driven by the re-allocation of risks and by other energy policies, consequences for prices will be clear only after one full business cycle passes. In the short term, it is more relevant to assess performance indicators of specific parts of the value chain, both in terms of the capital and labour resources required and the quality of service achieved.

Electricity generation is the most costly element of the value chain. In markets characterised by overcapacity and slow demand growth, it is common for markets that now face competition to withdraw some of the excess capacity. Capacity is either de-commissioned entirely or mothballed so that it can be brought into operation again at a later date. In Sweden, the ratio of installed capacity to peak demand fell from 35% in the early 1990s to 20% at the beginning of 2000. An immediate implication of liberalisation is that invested capital is put to better use. In New South Wales, part of the liberalised Australian market, the existing generation capacity (mainly coal fired) produced almost 12% more electricity in 2001 than in 1997, the year before the market launch.

Other indicators, such as fuel efficiency, suggest that generation plants are operated more efficiently in liberalised markets. Fuel represents the main cost component in gas-fired generation plants and about one-third of the fuel costs for coal-fired generation plants (the fuel component is less important to nuclear energy costs; the investment component dominates). Fuel efficiency, in terms of the amount of fuel used per kWh produced, plays an important role in the total economic efficiency of a power plant. The fuel efficiency depends on how close a plant comes to operating at its optimum. Recent research examines the effects of market-based incentives on fuel efficiency in various US states, comparing data from plants where competition changed incentives against data from plants that maintained their status quo (*i.e.* no such restructuring initiative). The results show that fuel efficiency improved by 2% under competitive conditions. The study also suggests that private or public ownership did not play a role; built-in incentives seem to be the factor that matters most.[5]

Transmission assets are also being put to more dynamic and enhanced use, in part because liberalised electricity markets place greater focus on trade and exchange between jurisdictions. Smaller countries in particular reap important benefits of cross-border trade and co-operation. In addition, trade across country and regional borders improves utilisation of interconnection capacity and often even prompts investments to increase this capacity. Electricity trade between countries and regions has increased rapidly since the 1960s; within some specific regions, such as the Nordic countries, the rate of growth has been even stronger since liberalisation processes began.

But perhaps more important is the fact that liberalisation brought fundamental change to the character of trade. Development of trade between Nordic countries is a good example. Norway liberalised its electricity market in

5. *Bushnell & Wolfram, 2005.*

1992; Sweden liberalised and joined the Nordic power exchange (Nord Pool) in 1996 and western Denmark became a partner in July 1999. An examination of flow over interconnectors between western Denmark and these two Nordic neighbours reveals dramatic changes in hourly energy flow from January 1995 to January 2000 (Figure 2).

Figure 2

Flow over interconnectors between western Denmark and Norway, and western Denmark and Sweden

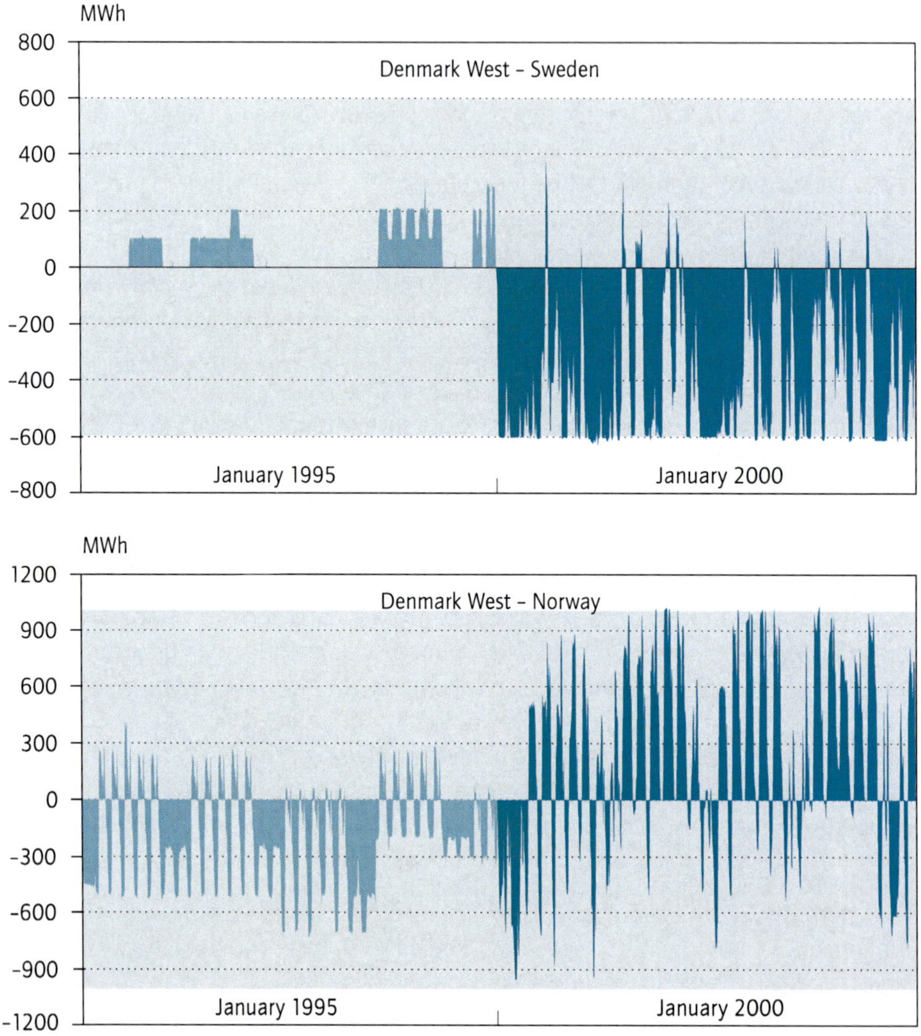

Positive numbers are import and negative numbers are export; the shaded area marks the rated capacity of the interconnectors.
Source: Energinet.dk

Evidence from western Denmark suggests that market players used interconnections more extensively in 2000 than in 1995 (Figure 2). Many factors influenced the size of the flow in both time periods, thus selected months do not necessarily show the full picture and cannot necessarily be attributed to the effect of liberalisation and competition. What is important is that the figures clearly illustrate change in the dynamics of use. In 1995, contracts directing the flow seemed more regular and, hence, more predictable - even several days, weeks and perhaps months in advance. In 2000, it is more apparent that flow can change significantly from one hour to the next illustrating that transmission assets are put to more dynamic and flexible use. Efficiency across the region is improving as a consequence, but it also stresses some new challenges of managing the transmission system securely.

Use of labour is another important productivity indicator. In this case, it is useful to compare trends in employment development. Electricity, gas and water sectors (Figure 3) is the sector category often used in standard statistical data compilations (data on Norway only include the electricity sector). The developments depicted may not all come from the electricity sector and some of the effects can result from out-sourcing various functions rather than representing a real reduction in the labour force.

Figure 3

Employment in electricity, gas and water (Index: 1990=100)

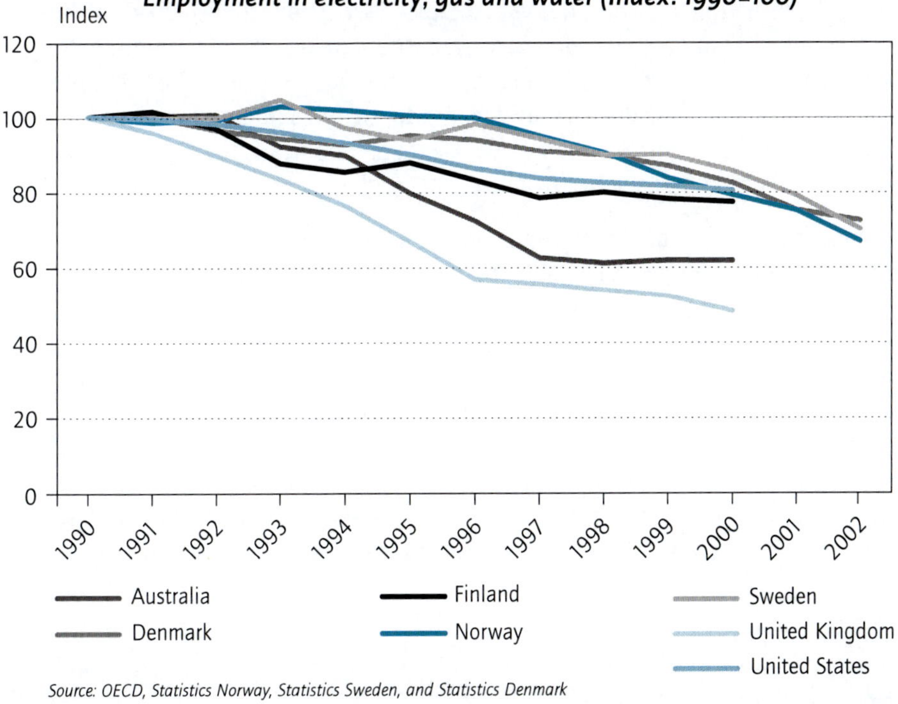

Source: OECD, Statistics Norway, Statistics Sweden, and Statistics Denmark

The data show a clear decreasing trend in the work force in all depicted countries. In the United Kingdom, the trend began right from the outset of the market opening in 1990 and is already evident in 1989 following adoption of the electricity act outlining market reforms (during previous years, employment was relatively constant). In Australia, employment decreases started in 1992, one year after the decision to reform the market and introduce competition was taken. In Norway, Sweden and Denmark, the decrease in employment really only started to pick up from 1999. This may reflect the fact that Finland entered the internal Nordic market in 1998 and Denmark entered in two steps in 1999 and 2000, thereby intensifying competition at the end of the 1990s. The trend underlines the important role played by the internal market for competition in Nordic countries. In Finland and the United States, the development seems to have been relatively smooth. In the United States, this could reflect the fact that the first step toward competition was introduced already in 1978 and that full-scale liberalisation was introduced in different states at varying points in time. This is supported by research suggesting that productivity of the US power industry was among the best in the world prior to liberalisation.[6]

A recent study on productivity in EU-15 member countries, conducted by the European Commission, concludes that labour productivity in electricity, gas and water sectors increased by 5.7% per annum between 1995 and 2001, up from 3.6% in the previous five-year period.[7] All in all, it has taken fewer individuals to successfully generate, transport and sell more and more electricity – at a time when electricity output was rising, in some cases significantly. This increased labour productivity has enhanced overall efficiency in the sector. But there have also been important social costs in the wake of this development.

Competition also improves transparency adding significant value in its own right. One side effect of the traditional vertically integrated model is that, in many countries, the electricity sector developed into a sector that serves many policy goals. As long as all costs could be passed on to consumers, collection of electricity tariffs often slowly evolved into thinly disguised tax collection. Revenues were used not only to cover the cost of generating and transporting electricity, but also to redistribute amongst different consumer groups and subsidise various activities. The net result is that the real costs of various policies become difficult to track, which may undermine economic efficiency – or at least obscure the efficiency of policy options. The parts of the sector now facing

6. *Joskow, 2003.*

7. *European Commission, 2004b.*

competition no longer have a guarantee that they can pass on all costs to consumers. In the new market, cross-subsidisation between consumer groups becomes difficult to manage and subsidies for other activities must be managed within the regulated network business. The experience from liberalised markets is that cross-subsidisation schemes become increasingly transparent and are often forced to end after a transition period. The costs of subsidies to other activities, such as support for renewable energy, also become more transparent. Information derived from increased transparency adds value in its own right; in many cases, it can lead to more effective energy policy adjustments.

It is one thing to produce a product with fewer resources - and quite another thing to do so without losing quality. One way of measuring quality is to monitor the amount of electrical energy never delivered to consumers. But quality of service, in terms of number and duration of interruptions, has rarely been reported systematically according to consistent definitions. The United Kingdom began systematically collecting such data in 2001, coinciding with the introduction of a quality of service incentive scheme for distribution network operators.[8] Initial figures show that quality of service - in terms of number of interruptions - improved 7% from 2002 to 2004. In Australia, legitimate concerns about quality of service arose when consumers - in some states, in some years - experienced more interruptions after liberalisation.[9]

However, there are many reasons to exercise caution in linking a quality indicator with the performance of liberalised markets. Final quality of service often depends on local networks: most instances of consumers being disconnected from the grid are due to mishaps in local distribution networks. Liberalisation has not changed local network activities. Networks are still regulated as in the past. Thus, quality of service is not really related to liberalisation except that liberalisation often prompts a greater focus on cost cutting in the economic regulation of local networks - which may ultimately lead to a lower quality of service. Finding ways to strike a balance between the short-term emphasis on cost-cutting without undermining quality in the longer run is a specific challenge for efficient, incentive-based economic regulation. Moreover, it is important to keep in mind that the effects of poor maintenance and replacement of networks will become evident only after a long time lag. Many markets are showing renewed interest in introducing specific quality parameters in their models for network regulation. For example, the United Kingdom, Norway and the Netherlands already introduced such elements in their regulation policies.

---

8. OFGEM, 2004.

9. ESAA, 2002 & 2004.

Various studies seek to assess the overall benefits of liberalisation. In 1997, one group of researchers studied the economic benefits of liberalisation, and the distribution of those benefits, in the England & Wales market.[10] They concluded that the total benefits correspond to some 5% of the total final electricity bill to consumers. The main source of the benefits was the shift of fuel from coal to natural gas. They also point out that the costs of restructuring were a significant element of the equation, which becomes particularly relevant when considering the costs and benefits of fundamentally changing a system as it happened in the England and Wales market in 2001.

Several studies in Australia asses the benefits realised by the NEM. *Securing Australia's Energy Future*, a white paper produced in 2004 by the Australian government, assesses the benefits from the NEM at AUD 1.5 billion annually.[11] In a major federal review of NEM, conducted in 2002, the Council of Australian Government's Energy Market Review suggested various changes and assessed that such changes would contribute *additional* benefits valued at some AUD 2 billion annually for the 2005-10 period.[12]

The Centre for the Advancement of Energy Markets (CAEM), a North American interest group, recently assessed the benefits of the internal PJM market, using cost reductions and efficiency improvements realised in the 1997 to 2002 period. In 2002, these benefits had a value of USD 3.2 billion, corresponding to roughly 15% of the electricity bill for that year. CAEM calculates the net present value of similar future benefits at USD 28.5 billion. In addition, benefits will be realised from improvements in dynamic efficiency that may arise over a longer time scale, resulting from new approaches to investment decisions, among other things.[13]

A recent OECD working paper summarised results of its work, *The benefits of liberalising product markets and reducing barriers to international trade and investment: The case of the United States and the European Union.*[14] The research used econometric and general equilibrium analysis to assess the annual benefits in terms of GDP gains per capita. Investigators assessed only static benefits from increased trade and better allocation of resources, even though it is recognised that dynamic gains from increased innovation may add significant benefits as well. The analysis covers several regulated sectors, including the electricity sector. In fact, the electricity sector was singled out as

10. *Newbery & Pollit, 1997.*
11. *Australian Government, 2004.*
12. *Council of Australian Government Energy Market Review, 2002.*
13. *CAEM, 2003.*
14. *OECD, 2005.*

the one in which the greatest government-imposed restrictions on trade and investment prevail - and, hence, the sector with the greatest potential for improvement. The paper uses Australia as the benchmark country for best practices in electricity. The total static benefits across all sectors are assessed to be 1% to 3% of GDP in the United States and 2% to 3.5% of GDP in the European Union, depending on the analytical approach applied.

There is value in comparing the performance of vertically integrated and regulated utilities with the performance of competitive electricity markets as electricity sectors operate today. It is perhaps more relevant, however, to assess the performance of liberalised electricity markets in the context of electricity sectors as they are expected to look in the future. The level of preparedness for future needs is not captured in the static indicators discussed above. It seems likely that future electricity systems will change character significantly. Liberalised markets place greater focus on risks and capital costs. As a result, there is increased demand for less capital-intensive and more flexible generation technologies that can be built in smaller, incremental steps. The rapid development of the combined cycle gas turbine (CCGT) technology fits these new needs well and is a good example of how the sector is changing.

At the same time, there is an increased array of energy policies designed to address environment and climate change challenges, which is already reinforcing a focus on smaller and more distributed technologies, such as wind power, and other distributed generation.[15] At the other end of the scale, there seems to be renewed interest in very large nuclear power plants, driven particularly by the future costs of greenhouse gas emissions. On the demand side, new technology and the new market framework enable consumers to take part in the decision making process. All in all, it seems likely that future electricity systems will be more diverse in terms of technology and unit size, and will include many small and distributed resources.

A future electricity system composed of a more diverse technology portfolio and exploiting the benefits of increased co-operation and trade calls for a very dynamic structure. The organisation of markets and system operation must give all resources an opportunity to interact with the electricity system in ways that ensure the best outcomes for consumers and owners of generation. This requires that all costs and benefits associated with each technology are allocated appropriately to create the correct incentives for a level playing field. It is likely that such a challenge for the future is best met by a market organisation based on incentives through competition. Allowing for interaction

15. The issue of distributed generation in liberalised markets is discussed thoroughly in an IEA study, IEA, 2002a.

amongst great numbers of market players. In a traditional vertically integrated and planned system, decisions are made centrally and any interaction from other players is possible only at the discretion of the utility itself.

Market Liberalisation is a Process rather than an Event

Prospects for establishing market institutions that effectively serve their purpose depend largely on the point of departure within each country. In Australia, there were no institutions upon which to build the National Electricity Market (NEM); it was necessary to create a new national transmission system operator, NEMMCO, as well as all the other required market institutions. The launches of markets in New South Wales and Victoria in 1994 were important steps and final market institutions benefited from these first experiences. Still, eight years passed between the final decision to liberalise in 1991 and the market opening in 1998. The Californian market was also built upon entirely new institutions. In the north eastern United States, co-operation within PJM was already well established prior to the market launch. In Norway, the market organisation was based on co-operation amongst Norwegian utilities, a mechanism that had been in operation since 1971. The England & Wales electricity pool utilised management systems already in place within the public utility, the Central Electricity Generating Board. On the whole, the Norwegian, PJM and England & Wales markets opened with much less time needed for preparation.

All of these markets underwent significant changes as they found themselves facing the challenges of new problems and new needs. In England & Wales, the entire market was fundamentally changed in 2001, eleven years after its initial launch. Australia undertook a major market review in 2002, which also led to a series of changes but no fundamental re-design. In the Nordic and PJM markets, problems and needs continue to arise at a constant pace, often stemming from issues related to co-operation between jurisdictions. There is an important lesson from this: the success of liberalisation does not depend on finding the perfect model from day one. Changes should be expected and appropriately managed. However, the costs of those changes, both for the market institutions themselves and for market players, become an important factor in further market development.

Increased market transparency is now recognised as a very strong instrument to ensure continuous market development towards more competition, improved market design and more effective markets. Prices that are publicly – and easily

- available immediately reveal indications of market power abuse. Transparency in managing transmission systems congestion often discloses deeper problems; transparency in real-time system operation improves understanding of the real costs of that service. With good basic market structures that ensure independent system operation and independent governance, pressures arising from transparency can lead to changes that improve market performance.

Analysing developments in the longest-running liberalised markets reveals various phases that are likely to be part of the liberalisation process. The first phase is preparatory: hard and careful political negotiations, legislation and market design are inevitable and even desirable. Technical infrastructure and management systems must be developed; institutions to operate them must be established. This phase can be shorter if structures that can be built upon already exist. A next phase is the development and maturing of the market. During the first years following market launch, some markets experienced problems with abuse of market power, which had to be addressed through changes to market design and legal action. Development of competition from new entries in the market is also common. And most markets also face some kind of crisis that ends up serving as a test of the robustness of the market. The consumer's role in determining the fundamental structure of the sector is being reinforced - slowly but steadily.

To date, no market can truly say it has passed beyond this transitional phase. It is likely that the next phase will be characterised by a continuing cycle of development, determined by the portfolio of generation plants. This might be called the 'investment' or 'business' cycle, during which markets may experience periods with excess capacity and periods with a tighter balance of supply and demand. It is also likely that there will be continuous pressure on market rules until a full business cycle passes (*i.e.* until old plants are closed and replaced with new ones). In reality, a full liberalisation process - from market launch to a robust and relatively stable market - will probably last at least one to two decades, or perhaps even as long as the economic lifetime of existing assets.

From the preceding discussion of liberalisation experiences and market indicators, it is clear that it is meaningless to think of electricity market liberalisation as an event. It is difficult - impossible, even - to take a snapshot and use that as proof of success or failure. It is also clear that the level of political involvement required does not end with the decision to liberalise or with the completion of the legislative work to establish the market framework. Monitoring, oversight and decision-making by governments and regulators will continue to be a necessary aspect in determining future development of the electricity sector. As they have in the past, market players will anticipate this factor and will react to the political, economic and regulatory risks it creates. Several examples show that at difficult junctures in market development a strong signal of political commitment can create the necessary market response.

In contrast, other examples demonstrate how small amounts of political intervention, which was only intended to be transitional, undermined trust in the long-term political commitment to market operation. One case in point is the temporary price cap on retail prices introduced in Ontario, which aimed to shield consumers from high prices caused by a tight balance between supply and demand. The intervention added uncertainty, which deferred investments and tightened the balance even further. As a consequence, the necessary market response of adding new generation - which likely would have solved the problem - did not take place.[16] Signals of willingness to intervene easily become self-fulfilling.

Distributing the Benefits of Liberalisation

Liberalisation of electricity markets is much more than a matter of just finding the right model. Most of the really difficult struggles in the political negotiations associated with the liberalisation process stem from dealing with the tensions that arise amongst different interest groups. Many of the inefficiencies in traditionally vertically integrated systems accrue from benefits (rents) or subsidies typically received by these different groups. Such rents create losses to societies, but the recipients of rents and subsidies unsurprisingly regard the potential loss of the rent as a real loss. When benefits are spread across large groups and losses are concentrated on small but powerful groups, the political landscape for negotiations becomes even more difficult. This landscape is different from country to country, depending on the point of departure and on the fundamental characteristics of each electricity system.

In the short term, the most obvious "losers" are those companies suddenly facing strong competitive pressure. This was not a big problem in countries where most utilities were state owned at the time of liberalisation. In England & Wales and in Australia, the competitive framework was relatively well known at the time utilities were privatised. The Nordic countries, New South Wales and Queensland, in Australia, still have large shares of public ownership. In many of the markets that are progressing at a slower pace, reluctance by incumbent utilities to effectively un-bundle system operations is an important barrier to development. In the United States, Spain and Denmark, efforts to recuperate

16. This example is described and discussed more thoroughly in an IEA publication on "Power Generation Investment in Electricity Markets. IEA, 2003a.

so-called "stranded" costs also played important roles. Companies argued that they had made investment decisions based on a regulated framework - decisions that will not be profitable in a competitive framework and for which they need to be compensated. In PJM, recovery of stranded costs was one of the key reasons for introducing a "capacity obligation" that enables incumbent generators to secure remuneration for all generation assets built prior to liberalisation (discussed further in chapter 5). Spain introduced a complicated compensation system, the Compensation for Competition Tariff, which links compensation to the price level in the market - and has effectively distorted market functioning.[17] Denmark compensated incumbent utilities through a lump sum paid over 10 years, funded directly by electricity consumers.

Various parts of the electricity sector workforce represent another powerful group that stands to lose from liberalisation - at least in the short run. To the extent that the sector has been overstaffed or that the change of the sector calls for a change in the staff, some employees stand to lose their jobs. Increasing dynamism in the sector is generally seen as profoundly threatening to most staff. As a natural consequence, labour unions and other workforce representatives rarely support the liberalisation process.

In many countries, different consumer groups also opposed electricity market liberalisation. The largest industrial consumers often have the most beneficial contracts with low rates as a part of industry policy. Residential consumers sometimes also receive preferential rates in the sense that even though they pay more than other groups, it is insufficient to cover the full costs of managing small customers. As a result, it is often the medium-sized industrial, commercial and service sector consumers that realise the greatest benefits from liberalisation. Large industrial consumers have often tried to argue for prolongation of preferential long-term contracts. In Norway, this segment of the market was offered preferential long-term contracts at the outset of liberalisation; these contracts are now maturing, which has triggered pressure on policy makers for renewals.

Cross-subsidies amongst different classes of consumers, depending on their willingness to pay for uninterrupted electricity, are another dimension that comes to light with unbundling and transparency. The vertically integrated sector's strong focus on the supply side may have been a preferable outcome for consumers who place very high value on uninterrupted and high quality electricity service. Often, for these consumers uninterrupted supply is critically important to their operations. In a vertically integrated system, such supply is

17. IEA, 2005b.

secured and the costs are shared by all consumers. In a liberalised market, consumers have the possibility to choose to consume and, therefore, also the possibility to choose not to consume if electricity prices rise above the value of the service it brings to them. This creates a situation in which consumers with the very high marginal value of electricity supply are left on their own to foot the bill. To date, these cross-subsidies have created only minor pressures but more firm reactions may be expected in the future, when the volatility of electricity prices increases.

In some countries in which electricity generation technology benefits from indigenous natural resources (*e.g.* hydro or cheap fuel), a large part of the benefits from liberalisation derives from better exploitation of this comparative advantage in an open market and may include re-distribution from consumers in general to owners of generation plants. Liberalisation and open markets enable local generators to sell electricity to a larger market and enable a larger market to benefit from a superior technology. This improves the economic outcome for the larger market but the benefits in the local market may be concentrated on the electricity generator. In Norway, preferential contracts offered in the past to the largest electricity consumers were based on the availability of cheap hydro resources. With the opening of markets, these resources can now instead be sold at higher prices in neighbouring markets to the net benefit of the Norwegian economy as a whole and to the benefit of Norwegian generators - but maybe at the loss of competitiveness for large electricity-intensive industries.

Many groups have important interests at stake in the liberalisation process. Pressure from these groups is a natural part of the political landscape and must be incorporated in the final package of solutions. However, if solutions to address social equity and re-distribution are allowed to distort the efficient functioning of a competitive market, a final positive outcome with net benefit for the economy may be jeopardised.

COMPETITION IS THE FUEL FOR EFFECTIVE MARKETS

The goal of electricity market liberalisation is to create benefits by introducing incentives for higher efficiency and more innovation. Effective incentives are created by introducing competition between market players. Competition exposes market players to the risk of losing market share, or even going bankrupt, if they are not sufficiently efficient and innovative. But it also provides rewards for taking risks and performing better than one's competitors. Failure to introduce effective competition can undermine the benefits of liberalisation in terms of lack of efficiency improvements and perhaps even deteriorating efficiency. In addition, abuse of market power has unacceptable consequences for financial distribution and equity.

The philosophy of *full* electricity market liberalisation is to introduce competition and choice in as many parts of the value chain as possible - from generation to consumption of electricity. Various countries have applied different steps to progress through the different phases of the liberalisation process. In some countries, a first natural step was to open opportunities for competition between independent power producers (IPPs) and incumbent utilities. The United States Public Utilities Regulatory Policy Act of 1978 (PURPA) is one such example. Other countries introduced the concept of contestability in various steps - that is, giving electricity consumers the freedom to choose their supplier in different steps over time, depending on their level of electricity consumption.

Competition can be introduced in most parts of the value chain, with a few exceptions. Local networks are natural monopolies and even though transmission networks could, in principle, be built in competition with generation, they are also still largely regulated monopolies. Transmission and distribution networks thereby remain, for the most part, regulated activities in which efficiency in terms of both investment and operation depends on the quality of economic regulation and the incentives it provides.

The traditional structure of the electricity utility is a vertically integrated entity, which is often state owned. Competition is introduced by breaking up that structure into its constituent parts and care must be taken to identify those sections that can actively participate in competition and those activities that should remain regulated. As the unbundling takes place, the communication and co-ordination within previously vertically integrated companies must be replaced with other structures to enable and improve the necessary incentive-driven co-ordination that leads to efficient outcomes. The breaking up of vertically integrated utilities and the establishment of new structures requires detailed structures for governance and solid legislative backing. Finally, when vertically integrated companies have been replaced with competitive markets, the level of competition must be monitored continuously.

Competitive Markets Replace Vertically Integrated Utilities

The first requisite to introduce competition is to separate or unbundle the natural monopoly network activities from all other activities. If an incumbent generator or a retail company maintains control over an affiliated network and can exclude or limit access to that network by competing generators or retailers, the network monopoly can be extended to an effective monopoly in the whole value chain. For this reason, transmission and distribution networks must be operated independently of generation and retail (Figure 4). While it might seem possible to slowly tinker with a vertically integrated industry in small steps, in fact, the very first step towards building a competitive market is the big step of deciding effectively to unbundled.

Figure 4

A new model for the electricity sector

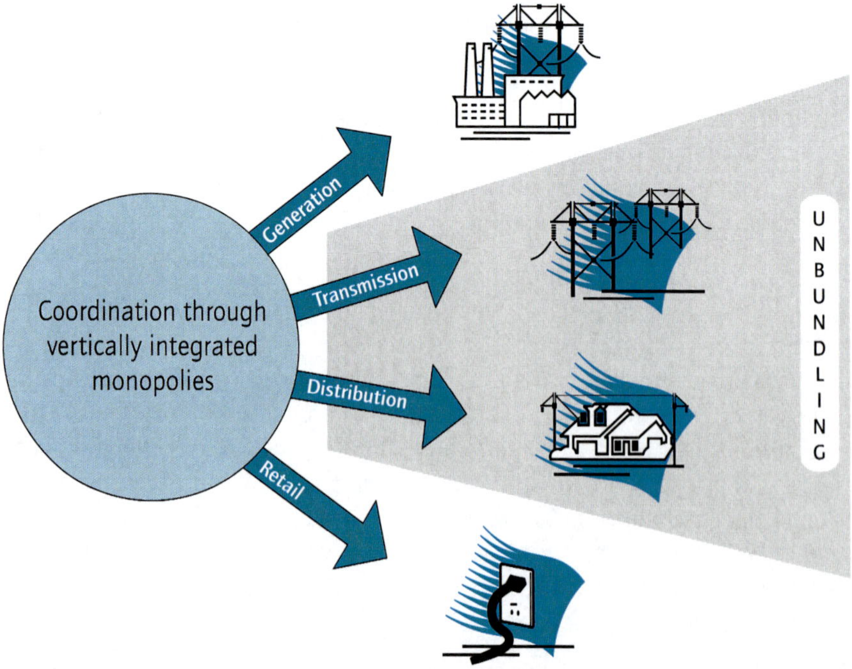

On the transmission level, the main challenge is to ensure that all generators have equal opportunity to feed into the transmission grid and all consumers share the same ability to extract electricity. Because electricity cannot be stored economically, maintaining quality of service require balancing of consumption and generation in every instant. At present, there is no cost-effective technology available to achieve this automatically. The alternative is to establish a system operator that operates the transmission system and thereby balances supply and demand in the whole electricity system. Centralised system operation is a necessary natural monopoly, which must also be fully independent of generation, trade and retail.

A central body responsible for system operation will manage the interface between the market and the actual physical outcomes. In fact, effective operation of the system must provide an indispensable part of the financial incentives in a liberalised market. By ensuring that the lights are kept on through balancing supply and demand in real time, a system operator delivers something resembling a public good. But the delivery of this public good relies on the rational actions by individual market players. They can only be expected to respond efficiently if their individual actions are committed in advance and thereafter metered to ensure fulfilment. Thus, the system operator must have full overview of generation and consumption, which can only be attained by establishing a very extensive set of rules regarding how generators, retailers, traders and consumers must interact with the electricity system. In the terminology of the European market directive, this is denoted as *regulated third-party access*. In other markets, such as the US and Australia, this is referred to as *open access*. Some of the most critical issues include rules for reporting and settlement of generation, trade and consumption schedules versus actual physical outcomes.

There are, however, a vast number of subtleties implied in secure operation of an electricity system - so many, in fact, that it is impossible to prescribe rules for every necessary interaction. Therefore, system operators still maintain certain discretionary powers, regardless of careful efforts to regulate grid access. In addition, many of the detailed rules could only be developed once systems had been tested in real operation and real problems emerged. It becomes critical that system operators have the right incentives to detect these problems and contribute to solving them in a way that enhances competition rather than limits it. This has created a particular challenge in many of the countries in which the system operator is unbundled from an old vertically integrated incumbent. The first set of rules are often based on old habits and principles, which were designed in the past to meet the needs of the incumbent - and will, therefore, often favour the incumbent, at least to a certain extent. To ensure his contribution to developing a competitive market, the system operator must have incentives to work toward improved rules that create a truly level playing field for all market players, with due consideration to the costs these changes may trigger.

Co-operation between neighbouring system operators is also critical to the development of trade between neighbouring jurisdictions. Again, such a development is only likely to be realised if the involved system operators all have incentives to co-operate. Negotiations about the basic rules and principles for cross-border trade can take place between policy makers and regulators in different jurisdictions. But it is the involved system operators who must find ways to manage the actual day-to-day co-operation. Incentives to work for the establishment of an effective market with a truly level playing field and effective co-operation across jurisdictional borders are only ensured by truly independent and effectively regulated system operators.

Liberalising markets have applied several approaches to the unbundling of transmission networks and system operations. While different markets have followed somewhat unique paths, each has had to address fundamental issues within two dimensions. One dimension is whether or not to keep transmission network ownership unified with system operation. The other concerns the form of unbundling in terms of ownership and level of independence.

Regarding the first dimension, North America and Australia chose to form independent system operators (ISOs), but to let previous owners maintain transmission. In Europe, most countries kept transmission ownership and system operation together in the form of transmission system operators (TSOs). Both solutions have a number of pros and cons. With ISOs, in which system operation and transmission are separate, it has been easier to create truly independent system operators that cover several jurisdictions. Critically, separate transmission ownership also makes it easier to develop a competitive framework for transmission investment, especially considering that transmission investments can be alternatives to investments in generation. Perhaps most importantly, full separation – in which roles and responsibilities are settled through legislation, rules and contracts – tends to improve transparency. The challenge here is to create governing and regulatory structures that ensure the required co-operation between system operators and transmission owners to achieve secure and efficient operation. It has proven difficult to effectively allocate all responsibilities, particularly when it comes to the more detailed aspects of system operation, with congestion management being an especially challenging issue.

In Europe, the argument for TSOs is based on the idea that transmission planning and ownership are integrated components of system and market operation. In this case, the challenge is to create governing and regulatory structures that provide transparency and efficiency in investment and operation and that do not favour transmission investment over generation investment and *vice versa*. In general, Europe has encountered slow progress in creating robust markets that support efficient cross-border trade; this could indicate that a second challenge has not yet been met - that of creating sufficient independence and incentives for committed market development among European TSOs. In short, the pros and cons for ISOs and TSOs are that ISOs enhance transparency but perhaps at the cost of co-ordination; TSOs enable co-ordination but perhaps at the cost of transparency.

Regardless of the approach used for system operation, it should be noted that transmission ownership is an important potential source for market power abuse. When transmission owners have large stakes in other parts of the value chain, incentives for investment in transmission may be distorted.

In the second dimension on ownership and the level of unbundling, the approaches taken by IEA member countries are even more diverse. The range includes examples of both private and state ownership, as well as various levels of unbundling. The European Commission issues annual benchmarking reports on the development of the internal European market and the fulfilment of the European market directive.[18] These reports introduce a terminology for levels of unbundling, classified according to ownership, legal separation, accounting or management. Unbundling by ownership and by legal separation are regarded by the Commission as being in accordance with the market directive.[19]

Table 1

Models for unbundling of transmission system operation

	Transmission System Operator (TSO)	Independent System Operator (ISO)	Transmission Owner (TO)
Unbundled by ownership - *state owned*	Denmark, Norway, Sweden	Australia	New South Wales (AUS), Queensland (AUS)
Unbundled by ownership - *privately owned*	Great Britain, Finland	Northeastern United States	South Australia, Victoria (AUS)
Legal or functional unbundling - *privately owned*			Northeastern United States

18. European Commission, 2005.

19. European Union, 2003a.

There are strengths and weaknesses in each of these approaches described above (Table 1), yet they share a common element: they have all managed to establish a necessary level of independence from other individual players in the market - an independence that is rarely challenged. In the United Kingdom and Australia, as well is in PJM and Nordic markets, all system operators are now unbundled by ownership. It could be questioned whether state-owned system operators are effectively unbundled when governments still maintain ownership over large electrical utilities operating in the competitive parts of the sector, as is the case in the Nordic market. An argument for state ownership could be to ensure that the system operator's main objective is to optimise social welfare. However, state ownership may actually make it more difficult to create clear incentives for efficiency and timely investment.[20]

With the establishment of independent system operators, the next challenge lies in ensuring necessary co-operation across jurisdictional borders to realise the full benefits of cross-border trade. Successful co-operation would also serve as an indicator of the level of independence of the co-operating system operators. Australian state and federal governments took the most radical approach to encouraging co-operation between system operators by forming one single, independent national system operator, NEMMCO. The PJM market has also developed according to the philosophy that market extension implies the expansion of unified system operation. System operators in the Nordic market are separate companies but have an association, Nordel, that serves as a forum to reach agreement on the rules directing trade and co-operation. Nordel has played an important role in supporting development of the Nordic market towards increased integration through harmonised rules. In Europe, system operators are organised through the European Transmission System Operators (ETSO), whose overall mission is to facilitate the development of the internal European market. Many of the operational aspects of physical co-operation in the synchronised continental European system are dealt with through the Union for the Co-ordination of the Transmission of Electricity (UCTE). In North America, the regional reliability councils that are members of the North American Reliability Council (NERC) play a similar role.

When it comes to unbundling local distribution networks, the main challenge lies in ensuring independence in relation to retailers. An effective system is needed to support smooth and easy consumer switching of retail supplier, to put all retail suppliers on a level playing field. The European Commission uses the same standard for unbundling of local distribution networks as for transmission networks (*i.e.* ownership, legal, functional or accounting). Only a

20. Considerations regarding governance and regulation of system operation are discussed in a recent IEA study reviewing electricity reform in Russia, (IEA, 2005a).

few countries, such as New Zealand, require ownership unbundling of local networks. Proposed legislation for ownership unbundling of local networks is also being discussed in the Dutch parliament. However, legal or functional unbundling is required in the Nordic countries, the United Kingdom, PJM states and the Australian NEM states and territories. In several countries that do not require ownership unbundling, discussions continue regarding alleged cross-subsidisation and limits to regulated third-party access on a level playing field. Several management considerations provide reason to challenge the independence of local networks and, in particular, the need to manage consumer switching, historic load data and consumers who chose to keep the initial supplier.

Unbundling is not enough to set the stage for a competitive market. While the general functions of the now-unbundled actors are well known, the detailed allocation of roles and responsibilities is critical to transforming an unbundled marketplace into a competitive one. The electricity system is operated through a number of action points and planning sequences over time, with due consideration to available resources, needs and constraints. These fundamental features do not change with unbundling and roles and responsibilities must be carefully distributed to ensure that all necessary actions will be taken in a timely manner. Some of the constraints, action points and planning sequences that are important in operating an electricity system are listed in Figure 5. There are several necessary planning sequences between the decision to invest in new generation, transmission capacity and fuel infrastructure and the final operation of the system and, hence, an ongoing need for co-ordination amongst stakeholders to ensure that all these steps are taken efficiently. Transparency and free flow of relevant information are critical instruments in enabling the necessary co-ordination.

Figure 5

Timeline for planning and operation

	15-2 years	1 year	1 day ahead	6-8 hours	15-1 minutes	Operation
Forecast	• Load • Generation • Transmission	• Load • Generation • Transmission	• Load • Generation • Transmission	• Load • Generation • Transmission	• Load • Generation • Transmission	
Contraints	• G & T capacity • Fuel	• Fuel • Revision & maintenance	• Working hours • Computing • Co-ordination with neighbors	• Plant start-up	• Ramp rates • Reserves • Co-ordination with neighbors	• Reserves • Black start capacity
Action	• Investment • Fuel contract	• Fuel purchase • Revision planning	• Day-ahead planning and optimisation	• Planning for operational reserves	• Balancing supply and demand	• Automatic regulation • Contingency management

As shown in Figure 5, the planning sequence starts several years in advance of operation. New generation plants and networks must be built to be available when needed and this must be planned as long in advance as it takes to design and build new equipment. In the shorter time frame fuel must be available for generation plants and existing equipment needs to be maintained: both activities must be arranged well in advance. Closer to the actual moment of operation, the system must start to prepare taking actual short term fundamental data into account. Electricity demand usually follows daily cycles that reflect the weather, the time of week and the season. Thus, all possible configurations of consumption and generation must be considered when analysing the transmission system's ability to transport electricity. It is imperative to conduct a careful analysis of system security and optimisation, which takes time to compute. In addition, staff is limited and staff who must be called to work outside of normal working hours are particularly costly. Given the daily cycle in consumption and the constraints associated with computing and staff, a day-ahead planning and optimisation cycle is the natural choice for most systems.

After the daily planning cycle, most decisions that will direct the bulk of the actions have been made. But many things can still change or go wrong. Resources must be available to manage any sudden changes in supply and demand. Many conventional thermal power plants take six to eight hours to start up from cold; nuclear power plants take longer and CCGTs can be operational more quickly. The start-up time often determines the last chance to respond to the need for system reserves. In minutes ahead of the actual moment of operation, supply and demand must be balanced; to safely perform this, it is necessary to consider the limits in ramping various generation technologies up and down. During actual operation the system must be prepared to respond to contingencies: to effectively deal with the unexpected, there must be reserves for frequency control and generators must have black-start capabilities.

This list of action points and time sequences illustrates some key points in the planning and operation of an electricity system but it is far from exhaustive. The important fact is that these key points do not change with unbundling. Thus, all responsibilities must be allocated appropriately between the system operator and market participants; the tools for delegating responsibility include legislation, regulation and market design. The next challenge is to establish an organisational infrastructure that facilitates all the necessary actions to fulfil the responsibilities, while at the same time respecting the constraints. It is particularly important that all stakeholders know the exact responsibilities of the system operator. This is usually limited to the real-time system operation and to dissemination of information when it comes to the longer term. All other activities on the longer term are left for the competitive market, relying on prices to provide the necessary signals to act on.

Allocation of responsibilities differs somewhat from country to country. In the Nordic countries, such deviations pose challenges for optimal co-operation within the increasingly integrated market. Within its ongoing work, Nordel describes four responsibilities for TSOs that are common in the legislation in all the Nordic countries: ensure the operational security of the power system; maintain the momentary balance between demand and supply; ensure and maintain adequacy of the transmission system in the long term; and enhance efficient functioning of the electricity market.[21]

It is one thing to allocate responsibilities and establish rules that enable all the necessary co-ordination and actions. None of this will work efficiently, however, without also putting in place the necessary incentives for individual market players. There must be incentives for competitive market players to invest and to operate plants efficiently. There must be incentives for all market participants to report planned generation and consumption as accurately as possible and in a timely fashion. Incentives can also be used to prompt delivery of all the other services that the system operator needs.

Establishing competitive markets for the full range of products necessary in electricity supply is the key measure to create an effective incentive-based framework. Organising a business by linking small companies together through contracts negotiated in open markets provides an alternative business model to operating within a large, vertically integrated company. Various economic factors often determine which of these two approaches is preferable within a given context. If large benefits can be realised by building large units and operating many units together, these economies of scale are best exploited in large companies. If the scale economies are less clear - and if the cost of establishing contracts that can mimic the structure of an integrated company are low - the market approach may be the best alternative. In short, low transaction costs make markets a possible alternative to vertically integrated companies.

Keeping transaction costs as low as possible is critical to electricity market liberalisation; one way to do that is to ensure that markets are well organised and transparent. The structures for communication within vertically integrated companies must be replaced with structures that allow for the same level of communication and interaction between independent market players and the system operator. If incentives are clear and competition is strong, good market structures will ensure that all players put their best effort into carrying out their actions efficiently, and actions will lead to overall efficiency even when decisions

21. *Nordel, 2005a.*

are taken on this decentralised level. This is potentially a much stronger, more efficient and flexible – and, thus, more robust – framework than can be replicated in vertically integrated systems. In such traditional systems, efficiency depends on management of individual utilities and on the ability to use economic regulation to mimic incentives for efficiency. In addition, vertically integrated structures motivated by regulated incentives will adapt to changing circumstances at a pace determined by regulation, whereas a competitive framework will effectively incentivise quick and dynamic adaptation.

One of the most important justifications for vertical integration is the need to manage risk. There may be benefits to gain from integrating several aspects of the market (*e.g.* generation, electricity retail, upstream fuel supply and natural gas retail) to manage the risk of the substantial capital commitments required. Hence, it is particularly important to replace the structures for risk management within vertically integrated companies; to create the opportunity for unbundled market players to enter into contracts, they must have the same possibility to manage risk. A liquid market for financial contracts offers a dynamic possibility for various players – generators, traders and retailers – to hedge risks at minimum transaction costs.

The retail side is an area in which clear economies of scale seem to remain in many markets. A relatively large number of customers seem necessary to make a profitable business of supplying small commercial and residential consumers. This is a logical consequence as long as product differentiation in contracts with residential customers adds only little value to consumers and as long as there are barriers to switching retailer. In that case, a retail business offers only a very standardised product with low profit margins. The lack of innovation and product differentiation could, however, also be an indication of lacking competitive pressure, from which retail companies would be forced to offer more differentiated products. More effective competition in the retail segment is likely to provide incentives for increased innovation, product development and differentiation – all of which might enable smaller innovative retail companies to operate profitably. (These issues will be discussed further, particularly in Chapter 4.)

Several markets, particularly the market in Great Britain, recently experienced varying degrees of vertical *re*-integration, *i.e.* generation companies have bought retail companies, or *vice versa*. In Australia, a retail company called Origin Energy moved to re-integrate by building new generation plants at various locations in the NEM. This re-integration raises concerns about the factors that drive such a development and the implications for competition. At present, there is no clear return towards vertical re-integration in the Nordic countries. This may be explained by a variety of reasons but it is likely that the relatively well-developed and liquid financial market is one important factor.

Some countries have taken a more voluntary approach to electricity market liberalisation, *i.e.* they let participants form the markets that they find necessary, without any detailed market design or formal rules established through a formal governance process backed by legislation. As all the functions of the value chain (from generation to consumption) must be maintained at all times, such an approach easily puts the incumbent in an advantageous position. Also, it is evident that unless the system operator has incentives to form and facilitate markets, the process develops very slowly.

In Great Britain, Australia, PJM and the Nordic markets, system operators all hold the formal responsibility for operating real-time markets. PJM also operates a day-ahead market and the Nordic TSOs own the Nordic power exchange (Nord Pool Spot), which operates the Nordic day-ahead market. (It should be noted that Nord Pool Spot is separate from the Nordic TSOs but they interact very closely through information sharing.) Similarly, TSOs in these four markets have the formal responsibility to operate markets for reserves and ancillary services. All in all, the term *system and market operation* better reflects the real responsibilities and activities of these companies and the establishment of markets is the result of active regulatory initiatives. There are also examples of countries that have opted for a clear distinction between system and market operation. In Spain, the electricity exchange Operadora del Mercado Español de Electricidad (OMEL) operates the market and the system operator Red Electrica de España (REE) operates the system, even though very close interaction and co-operation between these two companies is required to make this approach work.

Regulated transmission networks also play key roles in the market, even though they will not necessarily be incentivised efficiently through liberalisation. One of the challenges facing the regulatory framework is to find ways to incentivise transmission owners to maximise available transmission capacity *e.g.* when planning maintenance outages.

Legislative and Regulatory Framework for Effective Competition

Active legislation, regulation and market design, established collaboratively by governments, independent regulators and independent system operators, play critical roles in the development of liberalised and competitive markets. Liberalisation requires the necessary legal framework and a targeted process, launched by active government decisions. The intentions of government, as expressed in the legislation, then need to be implemented in a way that stakeholders can both predict and challenge. This requires a regulatory body that is independent of government. Many of the detailed market rules will directly influence the system operator's abilities to operate the system securely. Thus, system operators often play an important role in establishing market rules, either within collaboration with the designated authority or as an important advisory capacity. The roles of different actors and the instruments they have used to create the framework for competition differ significantly from country to country.

Government legislation to implement liberalisation ranges from relatively light legislation to a more detailed legislative framework. In Australia and United States, legislative development was strongly influenced by the fact that energy matters are, by-and-large, under the jurisdiction of state governments.

In the United States, the Federal Energy Regulatory Commission (FERC) has jurisdiction over matters regarding trade across state borders. Apart from the 1978 federal energy act, the Public Utility Regulatory Policies Act (PURPA), which introduced the first level of competition, two additional pieces of legislation promote the development of liberalised markets: the Energy Policy Act of 1992 and the recent Energy Act of 2005. The Energy Policy Act of 1992 gave FERC the authority to order open access for wholesale transactions between utilities. FERC tried to promote further development of liberalised markets with this authority but achieving fully liberalised and competitive markets required additional state legislation. In the northeastern states, in California and in Texas, liberalisation brought forward through state legislation ordered partial or full retail access for consumers to switch supplier. In the United States, a large part of the framework for operating the electricity sector is decided by the state Public Utility Commissions and FERC. Within the PJM area, it is PJM that decides on all the actual market rules and design features that, in the end, direct electricity flow – but always within the framework of rules and regulations set by regional and federal regulators. FERC also approves various entities to act as Regional Transmission Organisations (RTOs), whose role is to provide non-discriminatory wholesale electric transmission service under one tariff for a large geographic area. PJM is one of the RTOs designated by FERC.[22]

In Australia, the Council of Australian Governments (COAG) took a decision in 1991 to introduce competition in the electricity sector. This led to the necessary legislative reforms in the States and Territories, which hold constitutional

22. *Sally Hunt gives an excellent description of the development of the liberalisation process in the United States in* Making Competition Work in Electricity (Hunt, 2002).

jurisdiction regarding unbundling and open access. Following its decision to liberalise, COAG established a supervisory body – the National Grid Management Council, which worked in a comprehensive consultative process with governments, the electricity supply industry and electricity consumers. The Council's work led to the opening of the NEM in December 1998. The Australian Competition and Consumer Commission (ACCC; the federal body responsible for policing consumer and competition laws) authorised a new body, the National Electricity Code Administrator (NECA) to manage the electricity market code. Regulators in the country's six states and two territories were responsible for the regulatory framework of the electricity sector. COAG published a review of the NEM in 2002[23], which prompted the formation of a federal energy regulator covering both electricity and gas. In 2005, two new federal bodies, the Australian Energy Market Commission and the Australian Energy Regulators, replaced NECA and state regulatory bodies (some 13 bodies, in all).

Liberalisation in United Kingdom was triggered by the Electricity Act of 1989, which – in various steps – set the scene for privatisation, unbundling, third-party access to the grid, and retail contestability. The England & Wales market was reviewed in the late 1990s, which led to a major change in market design in 2001, also backed by the Utilities Act 2000. The British regulator, the Office of Gas and Electricity Markets (OFGEM), plays a key role in establishing market rules and developing the current market. In the original market set up, the system operator National Grid had a stronger role, building on the old structures for operating the system. The role of the regulator (then Office of Electricity Regulation - OFFER) was weaker and it was difficult for OFFER to introduce necessary changes. This was part of the motivation for the introduction of the new trading arrangements.

Legislation in the Nordic countries also created the framework for unbundling, regulated third-party access and retail contestability. In the Danish case, legislation effectively implemented the European Union market directives; Norwegian, Swedish and Finnish processes were ahead of the directives and essentially in compliance with them. The distribution of roles and responsibilities varies from country to country but, in general, Nordic regulators have the formal responsibility for market rules but focus on the regulation of networks; system operators play a critical role in establishing market rules and developing market design. This role has been reinforced through the co-operation within Nordel, primarily out of a need to further enhance efficiency of the integrated Nordic market through greater harmonisation.

23. *Council of Australian Governments, 2002.*

New Zealand presents another interesting example of governing structures. Here legislation provided the framework for privatisation, unbundling and retail contestability through the Electricity Companies Act of 1993 and the Electricity Industry Reform Act of 1998. Subsequent development of market rules was left to a framework based on voluntary agreement, which allowed for the development of a sophisticated market place. However, a general failure to further develop the market, as was necessary in the beginning of this decade, triggered the government to establish a regulator, the Electricity Commission in 2003. This effectively put an end to the concept of market governance based on voluntary agreement.

The rules and regulatory frameworks accommodate a variety of needs (Figure 5) and facilitate effective markets on three fronts: communication, market design and information. One set of very precise rules and regulations ensure smooth and effective communication between system operators, market operators, generators, retailers, consumers and traders. Additional rules concern scheduling and communication of bids. Rules related to market design and the market itself set the framework for trading and will, thus, direct all incentives. Rules on pricing principles and bidding requirements will direct all actions of virtually all players. Finally, transparency is a necessity for efficient markets, so there must be rules for disclosure of information. Market players, as well as system and market operators, must disclose the information needed to improve understanding of the fundamental conditions of the market. The Australian, PJM, British and Nordic markets all have very extensive sets of rules on all three of these key points.

The nature of liberalisation is that all markets are in a state of continuous development. The actual experience of market operation provides impetus for some developments, which are designed to improve overall market functioning. In several markets, incidences of market manipulation and lack of transparency prompted changes. Other developments result from changes in the underlying electricity system, such as interaction across larger and larger areas in United States and Europe or the increase in wind power in Denmark and the United Kingdom. The inherent lesson is that governing structures of an effective and robust market must be able to manage change. Market players, investors and policy makers all want stable rules and regulation; in fact, continuous changes are necessary to further develop markets and to adapt to changes in the framework. These framework conditions require a predictable and standardised system for managing change. User groups and close interaction between system and market operators and market players are important means of ensuring this predictability. Once this framework is in place, adaptability to change becomes one of the market's most important strengths.

There are some parts of the electricity sector that remain to be economically regulated after liberalisation, but with a reinforced focus on performance. System operation, transmission and distribution networks are more or less natural monopolies that will not be efficiently incentivised by competition. In principle, nothing has changed for these unbundled parts of the sector. They are monopolies as they were when being part of a vertically integrated framework, and their performance will, therefore, still depend on the incentives introduced by the regulatory framework.

Experience from well-developed liberalised markets shows a renewed focus on efficiency, particularly of local network companies and of new models for regulation that have been developed. Great Britain was the first to introduce a new performance based method with inbuilt incentives for improved efficiency, in a regulatory system using a revenue or price cap that is allowed to increase with inflation but also requiring continuous efficiency improvements of a certain percentage (known as the CPI-X or RPI-X method). Others maintained the standard method of allowing the network owner to pass on all costs to the consumer - plus a certain rate of return (the cost-plus method). After an initial phase of high focus on cost-cutting within the regulated businesses, there is now increased focus on quality. Great Britain, Norway and Sweden reformed their regulations to include service quality and reliability as specific elements of incentive packages; the general principle is to reduce revenues if quality does not fulfil a certain benchmark. Spain introduced a system in which network companies must compensate electricity consumers suffering from lack of quality.

Governing structures, in the form of legislation, rules and regulation, have been critical in establishing robust markets. However, there are also many examples that show the limits to these formal structures. In addition, the vast number of subtleties in system operation means that system operators will continue to have large discretionary powers: written rules simply cannot account for all possible situations. Regardless of the established framework, some severe situations may also create outcomes that are questioned and challenged for being unacceptable. There are several examples of electricity systems that faced extreme situations, which have often proven to be "the big tests" for markets. Extreme situations have always occurred from time to time, but probably less so in the past and with significantly less transparency. In traditional vertically integrated markets, it was easier to invest in excess generation and network capacity - and simply pass the full costs of such gold plating on to consumers.

From the launch of the Australian market in December 1998, there was a tight supply/demand balance in Victoria and South Australia. This led to very high peak prices, which raised political concerns over the functioning of the market. The South Australian government initiated an investigation to explore possibilities for intervention. The report concluded that intervention was not a good option and the government expressed its continued support for the market.[24] (This case is described in detail in Chapter 5.)

The Swedish electricity system faced a tight supply/demand balance during peak hours on two occasions in 2000 and 2001. During two particularly cold spells with extreme peak demand, there were fears of shortage. The situation prompted the government to request that the Swedish TSO, Svenska Kraftnät, submit a report outlining the situation and proposing solutions. Svenska Kraftnät stated that the market was intended to solve such situations and stressed the danger of government intervention. It also proposed various transitional measures that would have the least distorting effects on the market. The government followed the proposals and asked Svenska Kraftnät to implement the transitional measures.[25] (These measures are described in greater detail in Chapter 5.)

Another case in point is the Nordic market, which was struck by an extreme drought in the winter of 2002/03. Prices increased dramatically and the supply/demand balance was tight; fears arose of not having sufficient energy in the hydro reservoirs until they would start re-filling in the spring. The Norwegian government initiated an enquiry into the functioning of the market and presented its report to parliament, concluding that the market had handled the situation relatively well. Importantly, the Norwegian government showed continued faith in the ability of the liberalised market to deliver reliable, competitive supply.[26] (This case is described in detail in Chapter 3.)

In all of these cases, governments faced strong pressures to intervene because market opponents claimed the outcomes were unacceptable and ample proof of market failure. Government expression of its commitment to liberalised markets was likely critical to the healthy market responses observed in connection with these specific crises. More importantly, these signals of commitment provide credibility to the market for the future and thereby remove some of the political uncertainty that lingered in the minds of market participants. In 2002, the Ontario government intervened in the early stage of market liberalisation by capping retail prices. The intervention might appear to be of minor importance, but the signal of non-confidence it sent to the market

24. *South Australia Government Electricity Taskforce, 2001.*

25. *Svenska Kraftnät, 2002.*

26. *The Ministry of Petroleum and Energy, 2003.*

was critical. In fact, the Ontario Government came under strong pressure to implement a more severe intervention that essentially reversed the reform.[27] These examples demonstrate the importance of signals: the absence of signals of government commitment - or perhaps presence of signals of a willingness to intervene - may be self-fulfilling and actually deter efficient market response.

Regulating Competition

It is not enough to establish a formal framework that allows for competition through unbundling, regulated third-party access and retail competition. A second criterion for robust markets is the active participation of a number of market players, competing with each other. The actual level of competition remains a serious concern in many markets, due to high levels of market concentration.

It is illegal to exercise or abuse market power and competition authorities in most countries have means to address the breaking of laws. In legal terms, abuse of market power is, generally speaking, a matter of determining that a company with a dominant position is abusing its place in the market to realise substantial and sustained profit. In electricity markets, proof of abuse of market power is difficult to obtain. Part of the problem stems from the fact that it is often difficult to define the relevant market and the relevant product in a way that will hold up in the relevant legal forum. Is the relevant market the jurisdiction or the integrated market? Was the relevant electricity product traded in each separate half or whole hour or is it traded in a yearly contract? With such problems in clearly defining the relevant market and product, it becomes even more difficult to define the notion of substantial and sustained profit.

A related case came to light in Denmark in 2001. Danish market players complained to the Danish Competition Authority over alleged abuse of market power by the two dominating generation companies, Elsam and Energi E2, in 2000 and 2001. The Danish Competition Authority concluded it was not possible to prove that the two generation companies had abused their dominating positions. However, the both companies made an agreement with the Danish Competition Authority, committing to change their behaviour in the market. One element in this agreement was a commitment to act as a "market maker" that continuously provides bids to buy and offers to sell in the market within a certain price spread.[28]

27. *IEA, 2003a.*

28. *Danish Competition Authority, 2003.*

Both competition authorities and regulators play critical roles in the development of competition in several markets. While it is difficult *ex post* to prove abuse of a dominating position, there are better opportunities to regulate competition *ex ante* in connection with mergers and acquisitions. This has been proven true in several cases within the European markets, the largest of which required approval from the European Commission. A much-publicised case was the mergers of German utilities that led to the formation of the very large utilities E.ON and RWE. These mergers were approved by the European Commission and the German Bundeskartellamt, with some important conditions including divestitures and some changes to the market principles.[29]

In response to recent increases in prices, rigidities in cross-border trade and high market concentrations, the Competition Directorate and the Directorate for Transport and Energy of the European Commission decided to conduct a sector inquiry into electricity market competition. The inquiry will focus on the functioning of wholesale markets and how prices are formed, including levels of market integration and the functioning of cross-border trade. They will also focus on relations between network operators and their affiliates to examine barriers to entry in the electricity market.

Another instrument that has been used several times, and in various forms, is to require the sale of long-term contracts as an alternative to physical divestiture. In Europe, these contracts are called virtual power plants (VPPs); in other financial markets, similar products are known as option contracts. The buyer of the VPP obtains the right to draw electricity from a plant (or a pool of plants) according to the rules set in the contract. Contract forms vary but a common element is that the VPP is auctioned at a price that guarantees the right to draw energy at a predetermined energy price. The auctioned price corresponds to the option premium and the predetermined price corresponds to the strike price in the option contract. VPPs have been used in France, Belgium, the Netherlands and Denmark, in all cases as part of an agreement in connection with a merger or acquisition. VPPs increase liquidity in electricity markets and academic research broadly agrees that long-term contracts, such as VPPs, in electricity markets limit the possibility to abuse a dominating position.

VPPs are, however, far from a real physical divestiture. The rules for all VPP arrangements stipulate that buyers must give one day's advance notice if they wish to use the VPP. This excludes the use of the VPP from the real-time market for balancing power. VPPs are auctioned for a set period in the future: typically a month, a quarter or a year. If a dominating player has the power to raise the

29. *European Commission, 2000.*

wholesale market price, this will probably increase the value and the price of the VPP. It is thereby unclear how large a share of a monopoly rent a dominating player will lose in the end from this forced auction. Hence, it is unclear how effective this instrument is as a means to incentivise more competitive behaviour.

The easy access for new entrants is crucial to liberalised markets - especially those that have only a few competitors and for which there is insufficient political will or power to change that through forced divestitures. A dominating market player may have incentives to delay investment in new generation capacity: withholding capacity and investment will increase prices. In contrast, new entrants enhance competition and their role in adding new generation capacity to ensure that supply can meet demand may be even more important. Smooth, timely, clear and transparent approval procedures for construction of new power plants are essential for competition and reliability of supply. Easy access for new entrants provides a particularly good opportunity to enhance competition, especially in countries where electricity demand is growing at a high rate. Spain provides a good example; competition is not yet satisfactory but new entrants are starting to gain market shares.[30]

If competition cannot be developed sufficiently within a market, the most robust way to boost it is to extend the market by integrating several markets. The FERC strongly promotes this philosophy in its efforts to encourage the formation of Regional Transmission Organisations across the United States. In fact, this was the key philosophy in North America during the formation of northeastern markets, such as PJM. Market integration to enhance competition is also a key element in the philosophy of Australia's NEM. When Sweden decided to liberalise in 1996, one option considered was to establish a Swedish market that did not focus on market integration. This option was abandoned, in large part, because it was recognised that it would require breaking up of the large state-owned utility Vattenfall. The European Commission also sees market integration as the main path to a competitive European electricity market. This is underlined by the European Commission's TEN-E project, which identifies priority projects to relieve critical congestion areas and points in European electricity and gas networks.[31]

Regulation of competition has another complicating dimension in many countries. In many European countries, the largest utilities are state owned or have close ties with government through other connections. The largest utilities in New South Wales and Queensland are also state owned. These companies are often recognised as national champions and often provide

30. *IEA, 2005b.*

31. *European Commission, 2004a.*

substantial revenue streams; hence these companies often create conflicts of interest for governments. In the United Kingdom, Victoria and South Australia, governments privatised their vertically integrated, state-owned utilities during the initial phase of liberalisation.

Admittedly, it has been difficult for competition authorities *ex post* to prove abuse of a dominating position in day-to-day market trade. However, the threat of complaints and the force of publicity have been used with some success. The academic world, competition authorities and system operators have undertaken extensive modelling of market power abuse in North American, the United Kingdom, Australian and Nordic markets.

Competition indices are one of the analytical tools used to screen for potential market power abuse. The Herfindahl-Hirschman Index (HHI), which sums up the squares of market shares of individual companies, is the most well-known measure of market concentration. In electricity markets, HHI has proven to be a complex tool that should be used with caution. In some circumstances, it may be most relevant to compute HHI by looking at produced energy, but in most circumstances it is more relevant to look at market shares in terms of installed capacity. It may be even more relevant to analyse the specific types of technologies owned by various market players. A large nuclear unit must run as base load for technical and economical reasons. A CCGT plant is more flexible and may be a more powerful instrument in the hands of a dominant player. The annual *State of the Market Report,* produced by PJM's Market Monitoring Unit, is a very comprehensive market report that includes various HHIs as concentration measures, as well as analyses of the special circumstances in which one plant is a pivotal supplier.[32] Analysing market power for the pivotal or residual supply in specific market circumstances capture many of the complex considerations important in electricity market competition. Both PJM and Nord Pool have independent market monitoring units with responsibility of monitoring and analysing trade to detect breach of rules that support market manipulation.[33]

The Danish TSO, Energinet.dk, issues monthly market reports that contain data and brief analyses, including comments on abnormal price developments. Nordic TSOs and regulators co-operate to model market power on a continuous basis.[34]

Transparency is a prerequisite for analysis and understanding of the fundamental market conditions. All the necessary information must be easily available, in a

32. *PJM, 2005b.*

33. *Many aspects of analysing market power in electricity markets are discussed in Newbery et.al., 2004.*

34. *Nordic Competition Authorities, 2003.*

timely fashion, for all market players and authorities. Easy access to basic market prices is a first step. These are all publicly available on the Web sites of PJM, the Australian system operator NEMMCO, the British system operator National Grid, and Nord Pool in the Nordic market. Transparency on market data in the British market is limited by the fact that the new trading arrangements, implemented in 2001, are based on purely voluntary bi-lateral markets. In the Australian and Nordic markets, all information about fundamental factors in the market must be disclosed immediately. A market player who signs a contract with Nord Pool to participate in the market is obliged to give immediate notice of all changes to the status of generation plants above 50 MW. These changes are immediately posted on Nord Pool's Web site. Known as "Urgent Market Messages", these notices can relate to outages, plant re-connections, changes in schedules for planned outages etc. Without this information in the public domain, it is possible for utilities owning large plants to trade on information that only the owner possesses. There are several examples in the European market that illustrate how withholding information has given a plant owner an advantage.[35] In most financial markets such behaviour is called insider trading and is regarded as a serious criminal offence.

Efficient outcomes in electricity markets rely on timely investment in new assets and efficient operation of these assets. Skills in building and operating plants may, however, add little value if they are not backed with an understanding of what role these assets will play in the total electricity system. Information management and analysis is an equally important part of the value chain as traditional asset management. The level of transparency in markets can be seen as the level of information flow that enables the necessary information management and analysis. In most markets the skills to manage and analyse information develop into a new value-adding business segment – known as simply "trading", which was previously part of the vertically integrated utility although perhaps was not given the same weight. In the Nordic, British, Australian and PJM markets, traditional market players with generating assets or retail consumers have trading activities to manage these commitments and market positions. These markets also have market players who specialise in this segment, participating only in the trading role.

Traders play important – and, in some cases, pivotal – roles in making markets more competitive and efficient. But they can only participate in electricity markets if they are able to manage their risks. In that sense, the presence of traders is also a good measure of the level of transparency and liquidity in the market. In theory, a fully transparent market should not leave room for traders;

35. *Vilnes, 2005.*

some regard this as the ideal, considering the controversy the role of traders often creates. In practice, markets have proven to be far more dynamic and ever changing for this to be realistic and traders continue to provide crucial liquidity and competitive pressure.

Box 1 . Development of competition in the England and Wales electricity market

In the United Kingdom, the passage of the 1989 Electricity Act reflected the decision to unbundle and privatise the state-owned Central Electricity Generating Board (CEGB), which served England and Wales and to grant consumers above 1 MW open access to the grid. The first key steps of the restructuring took place over the next two years, with unbundling and launch of the trading arrangements in the so-called "Pool" in March 1990 and privatisation in February 1991. After unbundling, the sector in England and Wales consisted of a TSO, three generation companies and 12 regional electricity companies (RECs). The CEGB was divided into four bodies: National Power (50% of capacity), PowerGen (30%) and Nuclear Electric (20%) and the TSO, National Grid. All of the nuclear power plants were owned by Nuclear Electric. Originally, the intent was that the nuclear power plants should be part of National Power, which would then be large enough to absorb the business risks of owning and operating the nuclear stations. It turned out to be impossible to privatise National Power with these nuclear assets included in its portfolio, so they had to be withdrawn at a late stage in the process.[36]

From the opening of the market, Pool prices were lower than expected, primarily as a consequence of the structure of initial electricity supply contracts. Some of the factors in these contracts related to the obligation to buy domestic coal and the possibility to pass on costs to captive customers. Over the course of 1992-93 prices started to increase. In 1993 the regulator, Office of Electricity Regulation (OFFER), signed an agreement with National Power to divest 6 GW to a competitor – under the threat that their behaviour may be referred to the Monopolies and Mergers Commission, the competition authorities. In 1995, a "golden share" clause in the ownership of the 12 RECs lapsed, igniting a takeover and merger phase. In 1998 National Power and Power Gen each divested an additional 4 GW in exchange for approval to buy into the RECs now on the market.

36. Evans & Green, 2005.

From 1991 to 1999, in what is known as a "dash for cash", some 17 195 MW of CCGT was added to the grid, making up 23% of total installed capacity. By 2003, CCGT accounted for 33% of the installed capacity in Great Britain and 37% of the generation. These factors – the agreed divestitures, the takeovers and mergers, the construction of new CCGTs and additional voluntary divestitures – changed the structure of the sector completely. In 1990, the three largest generators accounted for 91.3% of total installed capacity while National Power alone owned 45.5% of the capacity. In 2000, the same three companies accounted for less than 50% of the installed capacity[37]; by 2004, they held only 37%.

In 1997, the UK government asked OFFER to review the performance of the Pool. OFFER concluded in its review (published in July 1998) that prices did not fall as much as they should have, despite the increasing level of competition. Companies with a dominating position were still suspected of "gaming" the Pool, particularly that segment of the market relating to capacity payments.[38] A decision was made to change the trading arrangements in the market to further enhance competition. The Pool was replaced by the New Electricity Trading Arrangements (NETA) in 2001. NETA was a fundamental change of market design; the key feature for enhancing competition was to change the trading arrangements from the obligatory central dispatch in the Pool to decentralised bi-lateral trading through NETA. The only formal market in NETA is the balancing market, which determines prices through an auction with discriminatory pricing rather than uniform auction prices used by the Pool. NETA's discriminatory auction price means that accepted bids will be paid the bid price. This is fundamentally different from the pricing principles in most other electricity trading arrangements, in which prices are determined by the price in the marginal bid. Pay-as-bid pricing versus marginal pricing is discussed further in Chapter 4.

Considerable theoretical and empirical research has been conducted exploring the NETA's impact on the level of competition. One recent study asks "Why did British electricity prices fall after 1998?"[39] Pool prices did decrease markedly from the end of 2000. Using econometric analysis, the study concludes that prices fell due to improved competition and capacity factors, rather than the introduction of NETA. The study strongly suggests that NETA did not change the behaviour of market participants.

37. *Hunt, 2002.*

38. *See Annex 3 and Chapter 5 for further descriptions of the England and Wales capacity payment.*

39. *Evans & Green, 2005.*

The most recent development in the structure of the British electricity sector is that there has been a vertical re-integration: large generating companies are acquiring retail businesses. With such a development, the focus of much research is now more on the effects NETA may have with respect to market transparency, transaction costs and liquidity. The main argument for re-integration seems to be a need for hedging in the generation business. Generation companies want a secure market for their product; in that regard a retail business becomes a so-called physical hedge. A physical hedge of that kind is very static and implies considerable transaction costs in connection with selling and buying retail businesses. Hence, the development could indicate that it is difficult to create the much more dynamic and cost-effective alternative in the form of liquid financial contract markets. Even worse, and of concern to several competition authorities, it could also indicate that retail can still be regarded as a semi-monopoly, considering that many smaller consumers do not search for a cheaper supplier on a regular basis. This creates an opportunity for vertically integrated companies to use the retail business to cross-subsidise the generation business, once again shifting the risks to retail consumers.

PRICE SIGNALS ARE THE GLUE

Generators, traders and retailers come to a market place to trade and make contracts. They all have needs and obligations, which are fulfilled by committing trading partners - through contracts - to generate or consume a specific amount of electricity at a specific time in the future. Contract negotiations will spell out the conditions (time, place, volume etc.) and the compensation - the price. If the conditions and products are standardised and well defined, market players have only to agree on the price. Such a market is said to be effective; it keeps transaction costs at a minimum and price becomes the signal that directs all actions. In effect, the price becomes the glue that links the necessary decisions in an efficient manner and makes the sector work with as little friction as possible. The standardisation of electricity as a tradable product becomes the critical challenge for successful electricity market liberalisation, enabling the creation of a framework for transparent prices that signal real cost and the willingness to pay those costs. If the prices reflect the real values at stake and if incentives are right, the resulting actions will lead to efficient outcomes.

Creating a framework that generates cost-reflective price signals to which market players can easily respond carries many inherent challenges. First of all, it is crucial to capture the true drivers of electricity costs and benefits, such as timing, location and volume of the delivery. The next challenge is to organise a market place in which market players and system operators can communicate prices based on these characteristics as smoothly and easily as possible, even though there will be important trade-offs between the level of sophistication of the pricing principle and the transaction costs of managing that pricing principle. Finally, in many markets important benefits are realised through trade across country and regional borders; in fact, cross-border trade may also be critical for effective competition in a specific region. Cross-border trade offers an opportunity for enhanced benefits; but it is also a constraint due to the need for co-ordination and harmonisation across borders.

Prices to Reflect the Inherently Volatile Nature of Electricity

By its very nature, electricity is a tricky product. It cannot be stored efficiently, yet consumers expect it to be delivered to fulfil necessary services in a timely, stable and reliable manner. In addition, consumption fluctuates from minute to minute and hour to hour and the conditions for generating electricity can change at very short notice. Electricity consumption often changes with the weather and there is typically a daily cycle with very quick and marked variations between peak hours and off-peak hours. It is likely that these fluctuations will increase in the future, along with the increased use of air-conditioning and other electrical appliances in both residential and commercial sectors.

In addition, supply and transport capability can fluctuate at short notice. Generation plants and transmission lines can face sudden problems that force them out of operation. For large generation and transmission units, one single outage can have a large impact on the balance between supply and demand. All in all, these factors make electricity inherently volatile and the volatility is not expected to decrease in the future. In fact, fluctuation of supply is likely to become more pronounced with increased shares of intermittent resources such as wind power. This volatility is an inseparable part of the characteristics of the service; it is not related to the organisation of the sector. That being said, the sector must be organised and aligned to effectively minimise and manage this inherent volatility.

In order for market prices to provide incentives that lead to efficient outcomes, prices must reflect the fundamental factors of the traded product. For electricity, this implies that prices for products and services in the different parts of the value chain reflect immediate conditions in generating, transporting and consuming electricity. Thus, prices that reflect the inherent volatility of electricity should be expected to be correspondingly volatile. The boundaries of the volatility can go from very high prices to reflect a tight balance between supply and demand, to very low - or even negative prices - to reflect an excess of supply.

Product prices in any liberalised and competitive market reflect considerations about marginal costs of supply and marginal benefits from demand. If competing with other suppliers, a supplier will be willing to sell his last produced product at the cost of this last marginal product - that is, at a price based on marginal costs. Extensive consideration and numerous experiments have tried to identify how this is best accounted for in the trading arrangements of liberalised electricity markets. A pricing and trading principle based on marginal costs means that the last accepted bid sets the price for the whole market. Some argue that such marginal pricing enhances the possibility of gaming between dominating players and has the potential to create windfall profits. Generators and consumers have no incentives to disclose their true marginal costs and benefits unless forced to do it through competition.

One alternative considered, which seems to appeal to the philosophy used in fully regulated sectors, is to pay dispatched generators the amount they initially bid (often referred to as "pay-as-bid" or "discriminatory" pricing). The standard critique of this approach is that generators will be forced to try to bid as high as possible, but still low enough to be dispatched. This forces generators to focus more on the marginal costs of their competitors than their own marginal costs, which is likely to prompt higher prices in the end. Most competitive electricity markets are currently based on a marginal cost pricing principle. One important exception is the British balancing mechanism, which uses a discriminatory pricing principle in which dispatched generators are paid what they have bid.

It is important to consider how market prices relate to supply and demand in a competitive market with prices based on marginal bids (Figure 6). Some technologies, such as hydro and wind power, have very low short-run marginal costs; nuclear power often has relatively low marginal costs while other thermal power has higher marginal costs depending on the technology used. Short-run marginal costs, such as those that determine competitive prices (Figure 6), compensate only a fraction of the long-run marginal costs, however. For nuclear, hydro and wind power, by far the largest share of costs are investment costs. The overall profitability of these technologies is highly dependant upon prices determined by other technologies when these are needed to supply the demand. For example, a nuclear plant will only recover investment costs and profits during hours when the market price is determined by other resources with higher short-run marginal costs.

Hydro power usually has a very different relationship with market clearing prices than the very simple relationships depicted in Figure 6. For hydro power owners, the cost assessment is not based on marginal cost considerations. Instead it is much more a planning problem in which the real costs are the alternative costs of using water today rather than saving it and using it at a later stage. The size of the reservoir is the constraint. The assessed "water value" determines bids rather than the short-run marginal costs. Hydro power plays a critical role in the Nordic market, with very high shares of hydro power particularly in Norway and Sweden.

High prices - and even price spikes - occur when only very few generation resources in the system remain unutilised. These resources often have very high marginal costs. In addition, a generator with the last resource available has extensive market power and may consider bidding a price that is significantly higher than the short-run marginal cost. The profitability of a plant intended for peak load relies on payment in the very few hours of annual operation that compensates for both variable and fixed costs. Hence, plants intended for peak load rely on a certain scarcity rent to be profitable.

Figure 6

Supply and demand in liberalised electricity markets

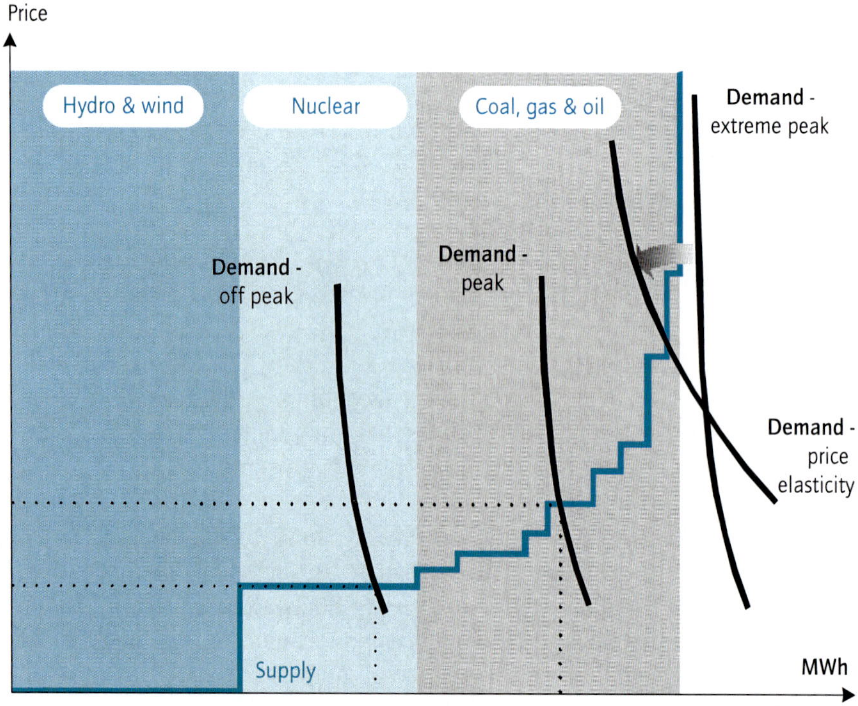

If one market player owns the last remaining generation capacity in a certain situation, the price may be determined by other constraints such as imports from neighbouring jurisdictions, a price cap or consumers who are willing and able to change consumption patterns.

The Australian, Nordic and British markets experienced short-lived but significant price spikes. In PJM there is a price cap of USD 1,000 /MWh, which is below the price spikes experienced in other markets. Price spikes and price caps have a crucial influence on investments, and demand participation (discussed more thoroughly in Chapter 5).

Excess of supply resulting in *negative* prices can occur when a generator wants to avoid closing down a plant, even if its marginal costs are not covered for a short period. This can happen for thermal plants (such as nuclear and lignite fuel plants) in which start up is costly, or with wind power in which there is lack of plant control and a subsidy or some form of obligation that acts as an important part of the effective payment. Hydro plants may also be willing to operate temporarily at negative prices, if the alternative involves bearing costs associated with spilling water. The Australian market has experienced negative prices for a few hours every year since market launch; prices in PJM market are also negative on a regular basis. In the Nordic market, negative prices are currently not allowed. However, Nord Pool announced that they will be introduced with new settlement software at the end of 2006.[40] Prices have been zero for numerous hours in the western Danish trading zone, where there is a high concentration of wind power. During these hours, prices would have been negative if the practice was allowed.

Electricity has a value to the consumer only if it is supplied at the right place, at the right time, in the right volume and at acceptable quality. The importance of these aspects is all directly related to the fact that electricity cannot be economically stored for later use. Quality is a standardised aspect of the service, ensured by the system operator. So far, quality is not an aspect that can be traded in the same way that the three other aspects are priced and traded.

Volume and timing aspects are relatively straightforward. Ideally, electricity should be priced and traded for every Wh for every second. This is what matters to the consumer. On the other hand, only the largest consumers can be expected to show interest in participating in the wholesale market, so a certain degree of aggregation is practical for the wholesale market even if electricity consumption is measured by the Wh. Wholesale electricity prices are quoted by the MWh and minimum bids in wholesale markets are one or ten MWh.

A certain aggregation is also necessary and practical when it comes to the aspect of time. In the second-by-second time span, system operators need a certain amount of capacity for automatic regulation. This generation capacity adjusts production according to a technical signal rather than an economic signal. Automatic regulation is only for the "fine tuning" of the system and results in a negligible share of total energy supply. So-called "real-time" markets are used to balance the system according to bids and offers in merit order. Balancing according to this normal economic dispatch makes it possible to meet variations between scheduled/actual generation and consumption, with a reaction time of five to 15 minutes. In most cases, energy delivered in real-time market is still priced and measured per MWh. Half-hourly or hourly pricing captures the bulk of the volatility in electricity generation, transport and consumption. The Australian and British markets are based on half-hourly prices; the Nordic and PJM markets use hourly prices. Longer term standardised contracts are also common, with delivery referring to a future

40. *Nord Pool, 2005.*

day, week, month, quarter, season or year. These standardised products use the half-hourly or hourly prices as reference. In short, considerations about timing and volume standards are trade-offs between capturing the fundamental factors that drive costs and benefits and the transaction costs of managing the standardised product.

In some markets, final settlement with market players includes an extra penalty for imbalances. The system operator must consider the final, aggregate, net deviation from the scheduled generation and consumption. Most market players will have deviations that contribute negatively to the final net-deviation, but there will also be a few players whose deviations contribute positively to total system balance. In Australia, PJM and Norway, those players that cause the net imbalance pay to those that relieve it, regardless of whether they relieve it by chance or because they were called upon by the system operator in the real-time market. In the British, Swedish, Finnish and Danish markets, those that contribute negatively to overall imbalance pay but there is no reward for those that contributed positively by chance. This dual-pricing system aims to provide extra incentive to reduce imbalances, assuming that market players will not rationally weigh costs of imbalances with the costs of improving forecasting. One side effect is that it also creates extra incentive to be large and thereby reduce imbalances through aggregation. Aggregation does not contribute to reducing system imbalance *per se,* hence the effectiveness of the extra incentive for improved forecasting is questionable. Moreover, it undermines liquidity in this market segment and does become a threat to competition.

Locational Pricing

The locational aspect of electricity pricing is the most controversial issue in establishing and developing liberalised electricity markets. As in all other markets, location is linked with transport service and transport costs money. In electricity, transport costs are significant. IEA estimates that, over the period 2003-30, some 13% of investments in the electricity sector within OECD member countries will be in transmission and 32% in distribution networks. The remaining 55% will be in new generation capacity.[41] On top of the significant investment costs, the issue is further complicated by the fact that resistance in electricity networks creates losses, which also adds transportation costs.

Electricity follows the path of least resistance ignoring any flow path that may have been intended with contracts. On any given transmission line, the

41. *IEA, 2004b.*

resistance and losses from transmission increase with the load. None of these relations are linear or constant, which makes optimising dispatch in economic terms highly dynamic and complex. The level of complexity is lessened somewhat in networks that connect single lines in a radial system to support generation and load centres. In contrast, complexity increases in meshed networks in which transmission lines criss-cross the transmission system, thereby creating several alternative flow paths. As an example of such "loop flows", a contract between German and French market players will usually partially be delivered physically via Belgium. Prices that reflect the real costs of generating, transporting and consuming electricity must properly account for the costs of networks and network losses.

Assuming that generation and load are evenly distributed across a radial transmission system with no congestion points, any generator would incur the same transmission loss. Thus, the optimal dispatch in such a system will always be the generators with the lowest marginal costs. In reality, no electricity system looks or behaves like this, but loss of economic efficiency can be minimised in systems with a radial structure. In a highly meshed network, however, the notion of efficiency and optimality may lead to fundamentally different results. Dispatch of the generator with the lowest marginal costs may, at a certain point, trigger grid losses that out-weigh its competitive advantage, thereby making a generator with higher marginal costs a cheaper alternative for the entire system. Dispatch of a generator with the lowest marginal costs may also lead to congestion somewhere else in the network, blocking access for other relatively cheap generators. If these considerations are not properly taken into account, it may force a very expensive generator into the dispatch, making the final outcome more expensive and less efficient.

Ideally, electricity is priced for every location in the grid with due consideration to both transmission capacity and incurred grid losses. The pricing principle which is closest to this ideal is the so-called "nodal pricing principle", which is applied, for example, in New Zealand and PJM. The philosophy of nodal pricing is that every transformer station in the transmission grid is a node and each node is priced, taking into account transmission capacity and grid losses (determined using computed loss factors). Nodal pricing properly prices all flows and constraints, including loop flows; thus it ideally creates full transparency and efficient incentives for investment and operation. However, many arguments are used for not applying nodal pricing, particularly that it would atomise liquidity in small shares, enhance market power, and increase transaction costs. It also raises issues of equity and distribution of wealth across regions and between consumers and generators often being the determining factor, particularly for policy makers.

The alternative approach is to apply pricing principles that ignore the locational aspect of electricity markets, or that address it only on a zonal basis. Zonal pricing aims to identify the main congestion points or areas in the transmission grid and then use them to form a group of nodes with uniform pricing. The Australian and Nordic markets could be described as zonal pricing markets. Each Australian state in the NEM forms a zone, as does the Snowy Mountain hydro plants region. The Australian system operator uses a nodal basis to account for network losses, computing loss factors for different reference points in the network. Dispatched generators are responsible for loss compensation. In the Nordic market, Norway is divided into three zones, Denmark is divided into two zones, and Sweden and Finland constitute one zone each. In this market, the network owner handles network losses; it is the owner's responsibility to purchase the losses in the market and thereby the loss is appropriately reflected in the zonal price as a part of the demand. In the British market uniform balancing charges are applied for the entire system, even after the inclusion of Scotland in 2005. The pricing principle in the British market does not include any transparent locational signal.

Network tariffs are another means of addressing locational aspects of pricing electricity. Tariffs for using the network can reflect the network costs associated with consumption or generation at a certain place in the grid. Such signals will normally be established on an annual basis, which is far from the dynamic, hour-by-hour pricing typically used in nodal or zonal pricing schemes. Sweden uses such locational signals in network tariffs. A more powerful locational signal can be sent along with the tariff for grid connection. For example, PJM uses so-called "deep network charges" in which the tariff for grid connection reflects many of the costs and benefits to the whole network with connection of a specific asset. Such deep network charges send very powerful locational signals, but it may be difficult to identify and allocate costs effectively and to manage them over time – e.g., when new generation or consumption takes advantage of existing networks.

All markets experience congestion or other constraints that limit the possibility of operating the system according to normal merit order dispatch. Nodal pricing should reduce these situations but experience from PJM shows that there will be situations that force the system operator to dispatch out of merit order. In some areas, generators must operate to ensure system security, regardless of their position in the merit order. Zonal pricing schemes also apply out-of-merit order dispatch to relieve internal bottlenecks. If there is congestion within a zone, by definition it cannot be priced transparently nor can it be taken into account in settling market prices. As a result, the system operator may be forced to dispatch a more expensive generator on one side of the constraint and remove a cheaper generator at the other end. Such a decision comes at a cost for the system operator: both generators will want compensation for the out-of-merit order dispatch (this type of activity is also called "counter trading" or "re-dispatch" in some markets). Often the system operator can use alternative means to manage congestion within zones. Instead of costly dispatch out of merit order, it may be possible to relieve internal congestion by limiting the available transmission capacity between zones and jurisdictions. This saves costs for the system operator but undermines the benefits of trade between zones and jurisdictions.

Transmission capacity available for trade is a critical factor when incorporating the locational aspects into efficient electricity pricing. System operators might limit the transmission capacity available for trade at levels below the actual physical thermal capacity of the line. Often, it is a question of system security concerns: system operators continually model and analyse the system, assessing system security against a pre-determined set of criteria. When system analysis points to breaches of security standards, one option is to keep transmission capacity available for responses to contingencies. To a great extent, system operators use methods for analysis of system security similar to those applied prior to liberalisation. These methods are often overly conservative, they are rarely based on probabilities for critical events and they rarely exploit the information on costs and benefits that the competitive market framework can contribute. System operators are now under pressure to better align practices with the new market framework and to ensure maximisation of available transmission capacity. A recent IEA study discusses more thoroughly the issue of system security in liberalised markets.[42]

There is also increased pressure for better transparency in the process of determining available transmission capacity and, thereby, improved predictability for market players. The pressure for higher transparency is even greater when there is zonal or no locational pricing and the independence of system operators can be challenged. In the European market, the European Federation of Energy Traders (EFET) calls for greater transparency in the management of transmission capacity available for cross-border trade.[43]

A number of examples illustrate the fundamental nature of the locational aspect within electricity pricing while also demonstrating why it remains controversial and complex. Nodal pricing is the ideal reference. In fact, if the chosen option for the locational aspect is different from the principle of nodal pricing, price signals will be distorted to a certain degree. The level of

42. *IEA 2005c.*
43. *EFET, 2004.*

distortion depends on many factors but, in general, it will increase with the number of congestion points left un-priced. Often the options applied seem to be more influenced by political considerations - in terms of social equity and distribution - rather than the pros and cons related to market functioning and system operation. Non-transparent management of interconnections is a serious barrier to trade and competition.

Shifting bottlenecks to jurisdictional borders and non-transparent management of transmission capacity restrict competition, and thereby automatically challenges the independence of the system operator, particularly when the system operator has close ties with the incumbent "national champion".

When the electricity market in Texas first opened (August 2001), there was only one control area or zone for the entire market. The agreed upon protocol was to manage congestion through dispatch out of merit order and to use a tariff up-lift to spread the cost amongst users. The target cost frame was some USD 100 million annually. In fact, during the first month of operation, congestion management cost USD 137 million. A decision was taken to divide the market into five zones (effective March 2002). Since then, expenditures on congestion management within these zones have remained significant at USD 200 million to USD 300 million annually. PJM market was also initially launched as a single price market but high costs for congestion management triggered development towards a nodal pricing principle after the first year of operation.

Sweden is an expansive country with high concentration of load in the south and high concentration of hydro power plants in the north. It constitutes a single zone in the Nordic market. The TSO Svenska Kraftnät manages congestion within Sweden, typically through out-of-merit order dispatch or by limiting the transmission capacity available for trade across borders. This triggers pressure, particularly from Norway and Denmark, to introduce locational signals instead of limiting cross-border capacity, which they regard as a discriminatory practice. In 2003, the Association of Energy Retail Consumers (an association for Denmark's largest industrial consumers) sent a complaint to the European Commission.[44] It argued that Svenska Kraftnät discriminated against Danish electricity consumers by limiting the transmission capacity available for trade between Denmark and Sweden - with only 24 hours notice - due to internal Swedish bottlenecks. This practice allows consumers in the south of Sweden to access cheap hydro power from

44. *FSE, 2003.*

the north of Sweden and Norway at the expense of Danish electricity consumers. The debate is still on-going. In 2004, the Swedish Energy Markets Inspectorate (the Swedish regulator) published a report showing that all Nordic TSO apply the practice of shifting internal congestions to the country borders to a certain extent.

A few intra-state congestions points exist in the transmission systems within the Australian market. Management of the Tarong constraint (southern Queensland) has generated much debate, particularly at times when congestions have been effectively moved to the border between Queensland and New South Wales.

Even with nodal pricing, which aims to identify and price all possible constraints, creating efficient and transparent price signals remains problematic. In PJM, it is also sometimes necessary to opt for out-of-merit order dispatch for system security reasons, usually in locations where transmission networks are very congested. Such out-of-merit order dispatches incur a cost for PJM, which is spread across all users. However, the resulting locational marginal price (LMP) in the affected location is not set by the plant that was asked to run out of merit order. Such practice is one of the barriers to establishing price signals that provide efficient incentives for operation and investment.[45]

A recent study modelled the costs of not using locational pricing in the England & Wales market. When this electricity market is modelled with 13 nodes, rather than the current large single region, the study showed that pricing with 13 zones can raise total social welfare by what corresponds to 1.5% of total generator revenues. Moreover, the zonal approach would reduce vulnerability to market power and create better investment signals. However, it may also lead to regional gains and losses that may be politically sensitive.[46]

All locational pricing gives rise to revenues. Two neighbouring areas with different prices will see a net-benefit from trade. In general, these benefits will be on the consumer side in the high price area and on the generating side in the low price area. But distribution of the benefits will also depend on how the trade is organised. A first pre-requisite for realising the benefits is the act of linking the two areas with a transmission line (Figure 7). The owner and manager of this transmission line thereby becomes the "gate keeper" and acquires the potential power to collect the benefits.

---

45. *Joskow, 2004.*
46. *Green, 2004.*

Figure 7

Electricity trade between two areas

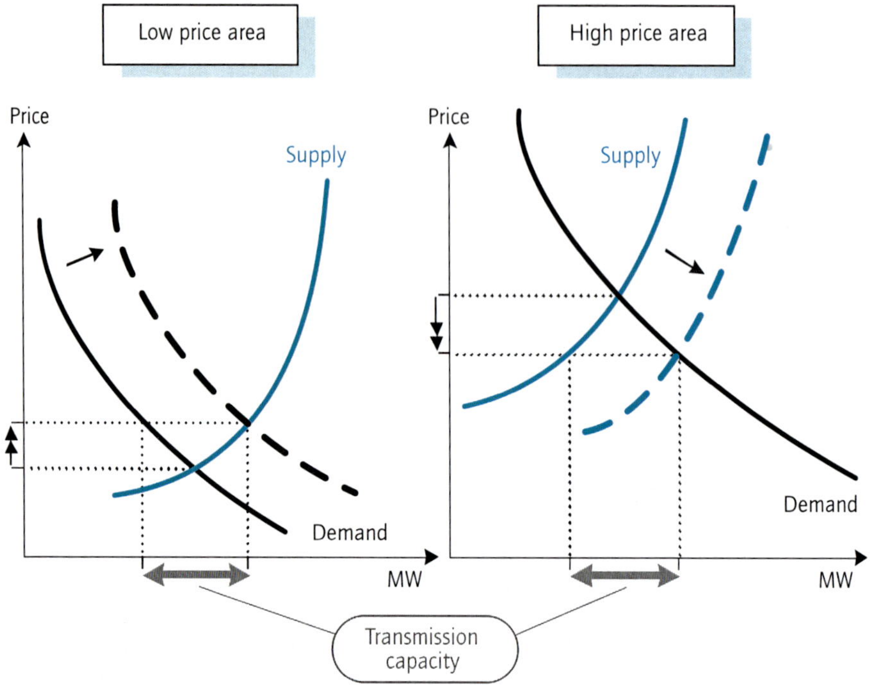

Transmission capacity linking two price areas or two nodes has an inherent value. The actor who controls flow over the transmission line will be able to collect what corresponds to the price difference between the two areas multiplied by the flow on the line. This stream of revenues goes by different names in different markets. In North America, and in theoretical literature, access to this stream of revenues is called a financial transmission right (FTR); in Australia it is known as the settlement residues while the Nordic region refers to congestion rents.

Access to FTRs can provide two key opportunities. Market players can use it as a hedge to manage risks associated with price differences between areas and it can provide a price signal for efficient investment in new transmission capacity. How these possibilities are exploited depends on the management of the FTRs. PJM initially allocates FTRs to transmission owners, holders of old contracts and builders of new transmission capacity. The "owners" can then use the FTRs to hedge bi-lateral trade and self-dispatch or for trading. PJM conducts monthly and yearly FTR auctions. In 2003, PJM introduced a new tool for allocating the initial FTR according to earned rights, *i.e.* instead of allocating the FTR that is specifically linked to a certain contract path, PJM allocates the annual revenue rights (ARRs) (which corresponds to the value FTRs sold for in the annual auction). An ARR holder can choose to transform it directly to corresponding FTRs if he wants to back a specific transaction or he can keep the ARR and perhaps buy other FTRs that better meet his specific needs. FTRs at PJM are firm, meaning that PJM guarantees the allocated capacity. If it turns out that the committed transmission capacity is not physically available, PJM must manage that shortfall through out-of-merit order dispatch.

In the Nordic market, TSOs grant Nord Pool a monopoly to all available transmission capacity between price zones for day-ahead trade. Thus, all FTRs are allocated to Nord Pool, who collects the congestion rent. The transmission capacity is firm, meaning that the transmission capacity made available for Nord Pool can be taken fully into account in the day-ahead settlement of spot prices. It is up to the TSOs to manage any deviations between the transmission capacity made available for trade and the actual physical transmission capacity, usually through out-of-merit order dispatch. Nord Pool allocates congestion rents to Nordic TSOs, who own Nord Pool, according to a formula. Nordic TSO's are bound by the European Union regulation on conditions for network access for cross-border exchanges in electricity according to which congestion rents can be used for three purposes: to guarantee actual availability of allocated capacity through out-of-merit order dispatch; to network investment; and as a part of the regulated income base, which must be taken into account in methods to regulate network tariffs.[47] Nord Pool also offers a financial product, known as a "contract for differences", that allows market players to hedge price differences between areas.

In Australia, NEMMCO controls the transmission capacity made available for trade by transmission owners. NEMMCO conducts quarterly auctions for settlement residues, resulting from price settlements based on the transmission capacity that is physically available at the moment of operation. In that sense, the settlement residue is not based on a firm transmission right. The auction of settlement residue provides an instrument for hedging risks of price differences but some of the value of the instrument is undermined by the lack of firmness. Settlement residues and auction revenues collected by NEMMCO are used to lower transmission use of service charges.

In the trading arrangements used in Australia, PJM and the Nordic market, transmission congestion is priced and managed simultaneously with the settlement of bids and offers from market players. The pricing of congested

47. European Union, 2003b.

transmission capacity is *implicit* in the settlement of market prices. An alternative approach is to make the auction of transmission capacity *explicit*. In reality, this is less efficient compared to the implicit and simultaneous pricing, which co-ordinates all aspects of the transaction. However, if a transmission line is not included in a uniform trading arrangement, the explicit approach can be a necessary, second-best solution. The German and Danish TSOs, E.ON Netz and Eltra, established an auction of transmission capacity over the Danish-German border, launched in 1999 when the western Danish system joined the Nord Pool trading arrangements. In the following years, other countries also established similar auctions along several other European borders, including the Netherlands-Germany and England-France borders. The European Commission's benchmarking reports indicate that both implicit and explicit auctions are characterised as market based and, therefore, compliant with the European Union market directive and regulation.

ETSO, the association of European TSOs, and EuroPEX, the association of European power exchanges, drafted a proposal for co-ordinating price settlement in neighbouring power exchanges. This would allow for co-ordinated action and thereby making congestion management implicit.[48] Simultaneous price settlement, in which transmission congestion between neighbouring electricity trading arrangements is taken into account, has been discussed in Europe for several years. Such co-ordinated action between neighbouring power exchanges goes by many names but in the most recent reports from ETSO it is called "market coupling". The proposal from ETSO and EuroPEX also includes a methodology to take into account – at least to a certain extent – loop flows in the highly meshed European transmission network, which is why they call the latest proposal "flow-based market coupling". The approach focuses on cross-border trade between jurisdictions but does not address the need for congestion management within countries and control areas.

Box 2 . Nodal pricing at work in PJM

PJM control areas contain highly meshed transmission networks. To meet the challenges inherent in meshed networks, PJM developed a trading arrangement based on locational marginal pricing (LMP), in which all relevant nodes are taken into account in the market settlement process. PJM serves 45 million people in 12 mainly northeastern states. Since its

48. ETSO & EuroPEX, 2004.

extensions in 2004, the entire PJM market area comprises 143 GW of installed capacity with a peak load of 115 GW. In that same year, PJM market participants generated, consumed and traded some 474 TWh of electricity.[49] *Currently, the LMP model includes 7 542 nodes (or buses), all of which are taken into consideration in price calculations and dispatch schedules. More than 6 000 nodes are with load; the rest are with generation. Buses (nodes) in the LMP model range from 765 kV to 4 kV level; almost half are higher than 100 kV, almost a third fall between 99 kV and 25 kV and the remainder are at lower voltage levels.*

LMP price calculations and dispatch schedules require input regarding thermal limits on all involved transmission lines - specifically, measures for the resistance on the lines, load bids and generation offers. PJM also requires information about start-up costs on all generators, which allow the LMP to account for the so-called "unit commitments". Basic function of PJM price settlement is as follows: all bids and offers for day-ahead trade must be submitted to PJM before noon on the day prior to operation. PJM then uses the LMP model to calculate prices and dispatch for every hour for the following day. The dispatch aims to minimise overall system costs while properly taking into account bids, offers, thermal limits and grid losses. Congested transmission lines lead to price differences between nodes. The confusion of managing congestion within such highly meshed systems, compared to congestion management in zonal pricing systems, is that all nodes in the network – not only the two nodes connected by a congested line – may, in principle, be priced at varying levels. That is, a congested line between two nodes may lead to price differences amongst several other nodes, reflecting the consequences of loop flows.

In real time, the LMP model runs every five minutes. The real-time market is accustomed to balancing any deviations from generation and consumption schedules. PJM charges imbalances with real-time LMP prices, i.e. *hourly prices that are integrated over all the five-minute intervals of each given hour. Both the day-ahead and real-time LMP models include hundreds of additional rules that try to accommodate the various needs and characteristics of generation and system management. This includes rules that blur the final price signal somewhat, such as a price cap of USD 1000 /MWh.*

49. PJM, 2005a.

There is no obligation to include all contracts in PJM's LMP market settlement. Self-dispatch is allowed to fulfil bi-lateral contracts or to meet own retail commitments. In 2004, PJM had a market share of 35% of total load in its own control areas. Self-dispatch and bi-lateral contracts are included as must-run offers in the LMP calculation. In that way they are not relying on PJM to be dispatched but they are still subject to locational prices that reflects the costs of transportation.

At the end of 2004, PJM comprised 34 control zones (divided into PJM mid-Atlantic Region and PJM Western Region), each covering a geographic area that was customarily served by a single utility. LMPs are based on a very complex model that accounts for the characteristics of more than 7 000 nodes, although the main congestion points in the market are concentrated on far fewer transformers, lines and interconnections. This facilitates the definition of trading hubs that comprise a number of important and normally un-congested nodes, and which ultimately serve as good reference prices. PJM calculates weighted average prices for three groups or hubs of specific nodes. The western hub is the most important reference point, with 111 nodes generating the largest share of liquidity in financial trading. There is also an eastern hub (237 nodes), as well as a hub comprising interfaces with neighbouring regions. Weighted average prices of several control zones are also used as references.

PJM measures congestion costs from being forced to dispatch out of merit order due to congested transmission lines. In 2004, the market's total congestion costs were USD 808 million, which corresponds to 9% of total PJM billing. The level of congestion can be measured in terms of number of hours during which a node is congested. In the same year, total congestion-event hours for the more than 7 000 nodes totalled some 11 205 hours; the most congested node experienced 1 784 congestion-event hours.

Several markets, including the Regional Transmission Organisations (RTOs) for New York (New York ISO), New England (New England ISO) and the mid-west (Mid-West ISO), use similar nodal LMP price and dispatch calculations for their own trading arrangements. These four LMP-based RTOs serve approximately 75 million consumers across 26 U.S. states.

Markets for Ancillary Services

Three elements define electricity as a product for wholesale trade: standardisation in volume (MWh), time (hour or half-hour) and location (node or zone). Standardisation results from trade-offs between capturing the fundamental factors that determine the value of electricity and the transaction costs of managing the trade. However, some important aspects of electricity cannot be captured by this standardisation. For example, an aggregation in time does not give the system operator all the instruments required to operate the system at a satisfactory quality. The system operator will need reserves for automatic regulation and frequency control in the case of outages, for safe operation in the very short run. Supply of reactive power is necessary to maintain voltage levels. The network also needs black start capabilities in case of blackout. Reserves may also be needed to balance the system on a minute-by-minute basis in case of more dramatic imbalances due to outages of large generation plants or transmission lines. Such operational reserves could be considered necessary for various reasons, including because it usually takes time to start up a plant and in integrated markets, the entire generation capacity can, in principle, already be reserved for domestic consumers and exports. The necessity of operational reserves in competitive and liquid markets is debated and is also argued to be an instrument that blurs signals that would trigger appropriate market response.

All these needs carry costs. The system operator must be prepared to pay for these services in order to maintain the incentives that prompt market players to supply them. Capacity reserved for ancillary services is not available for the ordinary market; thus, a capacity owner may incur a loss in supplying ancillary services rather than considering alternative sales in the market. At the same time, excessive payments for these services, particularly to dominating incumbent players, can distort the entire market. The approach taken in liberalised electricity markets is to define the needs for ancillary services as separate, specific products and to create mechanisms so that they can be bought by the system operator in a separate competitive market.

The Australian, British, PJM and Nordic markets are all in the process of developing specific markets for ancillary services. They are continuously adjusting products to include as many sellers as possible, without losing the necessary characteristics of the service. Australia realised substantial cost reductions through the establishment of a market for frequency control. The Norwegian TSO, Statnett, established an advanced trading platform for weekly purchase of operating reserves. These reserves are a right that enables the system operator to draw electricity under certain conditions; they resemble what is called an "option contract" in other markets. Statnett call their trading arrangement the "regulating capacity option market". Statnett boasts one marked achievement from careful standardisation through this option market: it has managed to attract a significant number of bids from the demand side. In Britain National Grid has also managed to attract substantial volumes of demand resources in its various market segments for ancillary services. Denmark opened market-based purchase of ancillary services in 2003 attracting a few new market players. Since the launch of the market-based mechanism, products have been constantly adjusted to attract smaller, distributed generators to the market. The Swedish and Finnish TSOs have taken a somewhat different approach. They own gas turbines, intended to serve as operating reserves. In addition, Finnish Fingrid decided to build an additional 100 MW as a means of contributing to required upgrades of operational reserves in response to the construction of a new nuclear power plant which re-enforces their approach of ownership rather than contracting in a market.

Cross-border Trade Creates Benefits

Open trade is one of the classical merits of liberalised and competitive markets. Open trade across country or regional borders allows both countries (or, indeed, several countries) to realise mutual economic benefits by finding and exploiting comparative advantages in the division of capital and labour. Electricity generation and transport include many factors that relate to resource endowments, geographical characteristics and national skills; it is also very capital-intensive businesses. Thus, there are many reasons to look for and exploit comparative advantages and many ways to realise large potential gains by optimising the use of assets across as large an area as possible. Clearly, for many countries, cross-border trade is an important source of realising benefits from electricity market liberalisation, particularly for smaller countries with geographically close neighbours.

In many situations, cross-border trade is also the easiest and quickest way to enhance competition. If it is not possible to create competition within a market, it may be achieved by enhancing the market itself. Again, the benefit is even larger for smaller countries in which economies of scale may limit the number of companies that can operate profitably.

The prospects of extending markets are, of course, limited by the availability of transmission capacity. Ideally, price signals on both sides of a congested transmission line should be used as a signal to invest in both generation assets *and* transmission assets. Transmission and generation assets are substitutes to a certain extent, reinforcing the notion that control over inter-connectors is one of the most powerful roles in an electricity market. The "gate keeper" in any given market acquires access to congestion rent and potentially controls the level of competition allowed. At any given level of transmission capacity, congestion rent will be reduced if transmission capacity increases. However, a capacity owner may not have the right incentives to invest - unless there is effective competition in the market. As a consequence, the independence of this "gate keeper" is essential to a competitive liberalised electricity market.

The business models for PJM and Australia separate system operation and transmission ownership. Independent system and market operators ensure that all congestion is priced and that transmission needs are transparent. There are two paths to transmission extensions. One follows competitive merchant lines, financed through congestion rents: electricity is bought by the transmission owner at the cheap end and sold at the expensive end. This model will be profitable only if a certain price difference can be maintained after the connection has been established. Optimal inter-connector capacity depends on finding a mechanism that just maintains a level of price separation capable of financing the line. The second pathway relies on enhancement through reliability requirements, using regulated tariffs to finance the extensions. Competitive merchant lines have not been successful thus far so most investment still rely on financing through regulated tariffs (see Chapter 5 for further discussion).

The transmission business model, used mainly in Europe, in which transmission ownership and system operation are kept together may offer better scope for co-ordinated planning of transmission lines to fulfil both reliability and trading requirements. However, with a monopoly on transmission ownership, private commercial incentives may not lead to efficient enhancement of transmission grids. Incentives will be distorted, as maximisation of congestion rents does not imply an optimal level of transmission capacity. Rather, transmission capacity is optimal when congestion rents just pay for the costs of the transmission line. Fear of distorted incentives is one of the main drivers behind the European Union's efforts to promote investments in new transmission lines relieving serious congestion points.[50]

TSOs in the Nordic market are moving towards a more holistic approach to transmission planning and investment. Nordel conducted a series of studies of the main congestion points and flow paths in the Nordic market, which identified five prioritised transmission projects. The Nordic TSOs agreed to proceed with a joint investment project worth EUR 1 billion.[51] The investment

50. *European Commission, 2004a.*

51. *Nordel, 2005b.*

package was developed with a reference to the net economic benefit of the entire Nordic market, rather than local gains and losses. The decision is backed by Nordic energy ministers but will be financed by grid users through tariffs (see Chapter 5 for further discussion).

In Europe the focus is not only on new investment. Only half of the 34 country-to-country interconnections between ETSOs 24 member-countries are allocated according to market-based principles.[52] Only the six Nordic cross-border interconnections are allocated according to the fully co-ordinated market coupling trading principle. Investment in new interconnections seems less important if existing cross-border capacity is not used efficiently to exploit the benefits of trade through fully dynamic trading arrangements.

Cross-border trade does more than create benefits through day-to-day trade. On a broader scale, trade and co-operation also create a more efficient, robust and secure system - particularly by joining forces to meet operational challenges through sharing reserves and other ancillary services. This joint effort to manage situations is particularly valuable to smaller systems, and particularly in circumstances of low probability but with serious consequences. Market integration in Australia and PJM also included integration of management of reserves and markets for ancillary services. In the Australian market, NEMMCO reduced the aggregate minimum reserve levels from 2233 MW in 2003 to 994 MW in 2004 - more than 50% reduction. This reduction is mainly attributed to the ability of inter-connectors to share reserves between regions and to differing demand patterns between regions.[53] Trade across jurisdictions has also reduced the aggregate need for reserves in PJM. In 2004, several steps were taken to extend PJM market area; summer peak demand increased 30% as a result of the extension. However, demand for spinning reserves (for which there is a market in PJM) increased only 20%, reflecting the value of increased co-ordination.

Electricity systems in Europe maintain a long tradition of co-operation and co-ordination regarding the use of reserves and other ancillary services, primarily through agreements within UCTE and Nordel. Markets for reserves and ancillary services continue to develop in all the Nordic countries and in several other European countries. There is one example of trade with reserves across borders: In 2003, Eltra, the western Danish TSO, bought operational reserves in Norway in agreement with the Norwegian TSO, Statnett. However, the increasing level of co-ordination still does not enable real reduction of reserves.

---

52. *ETSO, 2004.*

53. *NEMMCO, 2004.*

Box 3 . Nordic market: the Autumn the rain never came

In the autumn of 2002, the rain that usually falls in Nordic countries never came. Up until that point, it had been a wet spring and summer – particularly in Sweden. The picture began to change in early autumn as Sweden experienced unusually low levels of precipitation. A few months later, a similar picture emerged in Norway. In both countries, reservoir levels, measured as deviations from the normal level for each specific week, fell significantly (Figure 8). By the end of 2002, inflows to the Nordic hydro reservoirs in Norway, Sweden and Finland were 35 TWh below normal, which corresponds to 9% of total Nordic consumption in that year. (In Norway the situation was particularly bad: inflow was almost 18 TWh below normal corresponding to almost 15% of national consumption.) The droughts themselves were unusually severe in these three countries, but it was even more unusual that such severe droughts would actually coincide. A study referring to historic data assessed the probability of such severe droughts coinciding to 0.5%, or once every 200 years.[54]

In addition to the autumn drought, early winter was unusually cold and one of Sweden's largest nuclear power plants was out for refurbishment. The Nordic region, and Norway in particular, faced a severe energy crisis that had to be managed at every instant – even in the context of an uncertain future (i.e. without the clear overview that hindsight now offers). As indicated earlier, both Norway and Sweden were coming from a very wet spring and summer. As autumn progressed, it was valid to assume that the most likely future precipitation would be at normal levels. They did not normalise, and precipitation during the winter remained at very low levels. The first months of 2003 turned out to be relatively mild and the snow began melting relatively early, easing the immediate pressure to the system. However, Nordic hydro reservoirs only recovered to more normal levels in early 2005.

As a result of this unusual chain of events, electricity prices increased to unprecedented levels – culminating with a weekly average price of NOK 752 /MWh (EUR 104 /MWh) in the second week of 2003 (Figure 8). Nord Pool spot prices subsequently remained at relatively high levels through 2003 and 2004, and the market responded to the prices on several fronts. The most significant response to the energy shortage was substantial

54. ECON, 2003a.

production increases from thermal power plants in Sweden, Finland and Denmark. Several thousand MW of old, mainly oil-fuelled power plants were de-mothballed. Output from Nordic thermal power plants (oil, gas and coal) increased by 9 TWh in second half of 2002, compared to the same period in the previous year. In the first half of 2003, the increase doubled to 18 TWh compared with the same period of the previous year. Imports to the Nordic region from neighbouring countries were the second most important source to relieve the tight situation. In the second half of 2002, the net import to the Nordic region was 5 TWh; in the first half of 2003, it was 10 TWh. From June 2002 to July 2003, almost 10 TWh was imported from Russia alone.

A final significant response from the market came in the form of reduced consumption. According to a recent study, temperature-adjusted consumption decreased in Norway by roughly 3 TWh from October 2002 to February 2003, which corresponds to some 5%.[55] *In Sweden, temperature-adjusted consumption for the same period decreased by roughly 2 TWh. In Norway, the main reductions were from large industrial consumers, but households and electrical boilers also contributed. Electrical boilers have traditionally a large share of the heating market in Norway, particularly within process heating for industry. However, many of these electrical boilers can switch to alternative fuels, making this market one of the most straightforward fuel-switching opportunities in the country, even though only a part of the potential seems to have responded to price increases. The reaction from large industrial consumers in Norway was also influenced by a general economic recession. In Sweden, the main reaction also came from large industrial consumers. Marked differences in household responses in the two countries - Norwegians reacted by reducing consumption, Swedes did not alter behaviours - can be quite easily explained. In Sweden, one- to two-year contracts are the norm for residential electricity consumers; thus, price increases in the wholesale market had no immediate impact on household bills. In contrast, a large majority of Norwegian households hold short-term mark-to-market contracts that are adjusted according to spot prices; wholesale price changes can be reflected in individual bills on very short notice - as little as one to two weeks. (Consumer response during the Nordic drought is discussed further in Chapter 5).*

55. ECON, 2003b.

As the tightness of the Nordic energy balance intensified during the autumn of 2002 – and prices increased – the situation attracted more and more public attention. The situation awakened a broad Nordic debate about resource adequacy, diversification and market functioning. In Norway, the debate was broader and more intense than in the other Nordic countries, leading to significant pressure on the government to intervene. The high percentage of households with short-term contracts created a clear and fast linkage between current spot prices and prices charged to household consumers, thereby giving the debate a social dimension. One main point of criticism was that Norway continued to export significant amounts of electricity, even as the drought intensified during the last months of 2002. Effectively, Norwegian hydro producers emptied their scarce hydro resources to supply Swedish consumers. The development of the net export from Norway to Sweden, week by week, is shown in Figure 8, together with the weekly average spot price and the deviations from the normal reservoir levels in both countries.

Figure 8

Reservoir levels in Norway and Sweden, trade between Norway and Sweden, and Nord Pool spot price, 2002-2004

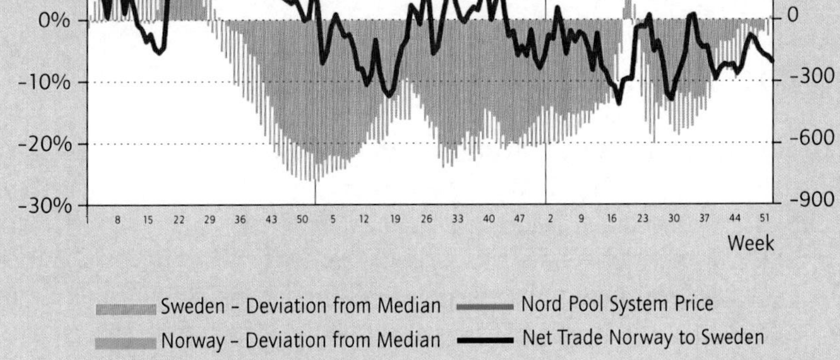

Source: Nordpool, adapted from ECON, 2003a.

Trade between Norway and Sweden is affected by many factors, particularly during winter when loads are high. Nevertheless, the relative *tightness of the hydro reservoirs seem to be an important driver for the flow between the two countries during this period. During October and November 2002, reservoir levels in Norway were low but the situation in Sweden was even tighter: Norway exported to Sweden. Despite public pressure to intervene to protect short-term national interests, governments allowed the market to allocate scarce resources according to market players' assessment of the value of water at any given point in time.*

The Nordic drought in 2002/03 put the Nordic electricity market under tremendous pressure. In that sense, it was an important test for the long-term stability of the market. The governing structures – from governments to regulators to TSOs – followed the development carefully but refrained from direct intervention. The market responded to the crisis by allocating scarce resources according to economic criteria and by attracting new resources from existing Nordic generation plants and through imports from neighbouring countries, as well as through demand reductions. The full range of potential opportunities seems to have been exploited to a great extent. The Norwegian government presented a report on its view of the situation and proposed future measures to the Norwegian Parliament in December 2003.[56] The report declares that the Norwegian government will continue to base its energy strategy on a well-functioning electricity market with trade between countries and concludes that the market managed the situation well. At the same time, it also recognises that attention must be given to the prospects for investments in new generation capacity such as renewables and gas.

Market Models

This book focuses on issues that are important to establishing competitive and robust electricity markets. PJM, Australian, British and Nordic markets were chosen as focal points because of their relatively long and successful experiences. These markets are similar in the sense that they have all seriously addressed key parameters necessary to creating competition: unbundling; regulated third-party access; opening of retail markets; and establishment of comprehensive trading arrangements. Yet the market models or designs

56. *Ministry of Petroleum and Energy, 2003.*

chosen by each country vary significantly. While they all address basic aspects that determine the value of electricity – time, volume, location and quality – they apply different weights on the different aspects.

Much of the transmission network in the United States is highly meshed, including the network in PJM control areas. It is natural, and probably necessary, to apply a nodal pricing principle in such a market. In contrast, the Nordic and Australian markets have load and generation centres that are interconnected in more radial networks and have fewer loop flows. In such systems, the loss of efficiency that comes with applying a zonal pricing principle seems more acceptable; considering the increased complexity of a nodal system, zonal pricing appears to be a natural choice. That said, the inefficiencies that arise when pricing principles fail to transparently reflect the real congestion continue to prompt debate in both markets. The England & Wales electricity system is relatively well interconnected internally but has relatively weak interconnections with neighbouring systems. In this context, the efficiency losses inherent in a single price zone appear acceptable. However, integrating Scotland into the British Electricity Transmission and Trading Arrangements (BETTA) created more debate, primarily because the interface between Scotland and the rest of the market is a clear congestion point and a decision was taken that it would not be priced transparently. Debates concerning more fine-tuned, locational pricing in the Australian, Nordic and British markets appear to be heavily influenced by political concerns for regional re-distribution. Even if such social aspects can be managed through other policy instruments, they appear to get a higher weight than considerations for cost-reflective price signals and well-functioning markets.

Timing of trade also differs considerably from market to market. In the Australian and British markets, trade is concentrated close to real time; it is up to market players to make contracts that reach more than two to three hours into the future. Considering that both these are isolated (or in the British case, relatively isolated), there is no particular requirement for careful and time-consuming co-ordination with neighbouring systems. In Australia, the need for co-ordination was addressed very comprehensively by forming a joint system operator with responsibility for operating the system and market across all involved states and territories. In the absence of requirements for external co-ordination, it appears natural not to complicate things by introducing several official trading cycles. These choices, however, raise debate as to whether transaction costs in forward markets will become so high as to discourage participation of some market players. In Australia, attracting active demand-side participation in the market remains a serious challenge. In the British market, there seems to be a trend of vertical re-integration within the generation and retail segments of the market, indicating that transaction costs in forward markets are too high. In addition, operators of small plants, particularly intermittent wind generators, have complained about the market structure.

In PJM and Nordic markets, market integration and trade across jurisdictions has been a top priority from the start. Both the Nordic and the northeastern U.S. markets use day-ahead price settlement as the main reference point and both have long traditions for cross-border trade, so the need for co-ordination is a natural focal point. Real-time markets provide corrections to the day-ahead price settlement and longer term forward contract markets use the day-ahead price as the reference. Such practice allows for more careful co-ordination and planning well in advance of operation. Also, day-ahead market settlement has the advantage of opening the market to players who do not have the resources to maintain 24-hour market monitoring, seven days per week. This contributes to liquidity, competition and flexibility but may, however, remove some of the continuous alertness from market players and dull their responsiveness to sudden changes. It is also likely that such an organisation of the market requires relatively more operational reserves

Day-ahead commitments related to transmission capacity made available for trade create a framework for firm contracts for market players but potentially at the cost of loss in overall efficiency. In fact, day-ahead market prices may not correctly signal the real physical conditions of the system at the time of operation. It becomes critical also to operate an effective real-time market, including the option to trade across interconnections. PJM includes a careful real-time market settlement that also integrates and optimises flow across PJM control areas in real-time trading. Nordic TSOs operate an integrated real-time market for balancing services that ensures optimal flow across Nordic interconnections in real-time trading

Other physical characteristics of the different systems also play important roles in the development and success of efforts to organise market and trading arrangements. The Nordic market originates from the liberalisation of the Norwegian market, where almost the entire generation capacity is hydro based – mostly from reservoirs, which constitute a very flexible balancing resource. Sweden also has a substantial share of hydro power. In 2004, almost half the electricity generation in the Nordic countries was from hydro power, giving the whole Nordic market access to this flexible resource that effectively enables relatively cheap short-term balancing of deviations from day-ahead schedules. The Nordic market has developed substantially with the inclusion of other Nordic countries and, thereby, with the inclusion of larger and larger shares of thermal-based power plants. Thermal power plants are generally costly to start up, so the short-term marginal cost of operating for one hour is substantially higher than operating for *e.g.* 12 hours. As a consequence, Nord Pool introduced "block bids", which allow a generator to bid at a certain price under the condition that he can operate for a certain number of hours. The sophistication of these block bids has increased continuously. In PJM, such so-called "unit commitment" considerations are included directly into the market dispatch process.

There is a natural focus on cross-border trade in continental Europe. Most markets operate on a day-ahead planning and trading cycle, which facilitates co-ordination between TSOs and market operators. The ETSO-EuroPEX proposal on flow-based market coupling also builds on that by focusing on flows across borders between countries and jurisdictions. Considering the highly meshed nature of the European transmission system, which is similar to the transmission system in United States, it appears questionable whether a trading arrangement focusing on countries and jurisdictions will provide sufficient locational signals in the long term. To date, only limited efforts have been made to integrate real-time and balancing markets across Europe.

All markets are under continuous development. Trading arrangements in the British market underwent a major reform in 2001. The common drivers for change include considerations about market power, demand-side participation, intermittent resources, small distributed generation, risk management and transparency.

In summary, it is now evident that the design of an appropriate market model is a matter of creating trading arrangements that fit the specific circumstances of each electricity system while also addressing the key issues of unbundling, third-party access, cost-reflective pricing principles and transaction costs. It is also clear that there are important trade-offs between efficiency and complexity. There is no winning, "one-size-fits-all" market model. However, there are some common features in the pressures for development, particularly with respect to technology trends and the need for more demand participation. In that sense, there may be a certain level of convergence in market design.

RISK MANAGEMENT AND CONSUMER PROTECTION

Uncertainty breeds risks. This is certainly true of the considerable uncertainty connected with many fundamental factors that determine electricity generation, transport and consumption. But it is also true that uncertainty - and thereby risk - can be reduced to some degree through skilful operation, analysis and information management. At the same time, a number of factors connected to energy generation will always leave some uncertainty: weather, technology development, fuel prices, competition, etc.

Uncertainty can also be found in the regulatory framework. Here, it is in the hands of regulators and policy makers and it is important that these two stakeholders consider the relationship between regulatory uncertainties, business risks and costs. At the same time, a commitment to remove such uncertainty entirely is almost certainly not credible. The bottom line is that in the electricity business - as with any other business - risk will always be involved in every part of the value chain. For market players, it is important to minimise risks through competence and manage the remaining real risks to align their appetite for risks compared to the rewards.

Efficiency requires that all elements are properly taken into account in all decisions, including the fact that information will always be imperfect - there will be uncertainty. The framework of incentives should be structured such that risks are allocated to those who make decisions and who hold responsibility for taking uncertainty into account. This is not the case in a vertically integrated and regulated sector in which all costs can be passed on to consumers. In such a system the consumer typically bears all the risks but investment and operational decisions are made within the vertically integrated utility, although the decisions may be subject to regulatory approval. Liberalised markets allocate risks to the real decision makers; utilities face the risks directly and must take them into account when making decisions. Market liberalisation does not change the risks or uncertainties in the electricity business. However, certain business risks may appear new to utilities and financial institutions lending money to these utilities. In fact, the risks are simply more transparent and appropriately shifted from consumers to the relevant market players.

The challenge of managing risks in liberalised electricity markets is multi-faceted. Electricity is inherently volatile with large variations both from hour-to-hour, day-to-day and season-to-season. At the same time, most assets in electricity have economic lifetimes of 25 to 60 years. In the past, traditional cost considerations determined most aspects of investment decisions including choice of technology. Fixed and variable costs were assessed for the economic lifetime of the asset and discounted back to present value, applying a discount rate that reflected the time-value of money – *i.e.* the levelised lifetime cost approach. This approach created a situation in which there were no or very small risks for decision makers, and low discount rates were applied to reflect that. This picture changes dramatically when risks must be fully considered in liberalised electricity markets; there is a greater appreciation of technologies that can be developed and implemented in small, incremental steps with low, up-front capital costs – especially where demand growth is low, irregular or otherwise uncertain. For example combined cycle gas turbines (CCGT) have many advantageous attributes for liberalised markets and are often a preferred technology.

In electricity markets, risks are managed through contracts – contracts that connect generators with consumers. Electricity is physically delivered only in the actual moment of operation: a generator and a consumer/retail company make the final physical commitment when a schedule is forwarded to the system operator or the system operator makes the final dispatch. In the minutes, hours, days and even years preceding this moment of physical exchange, market players can change their commitments many times by trading contracts. Thus, it is easier to conduct business through financial contracts – *i.e.* agreements to exchange money as opposed to agreements to exchange the actual physical product. A generator and a consumer will establish a contract on electricity supply during a specific period in the future and will agree on a specific price. If the market price turns out to be lower than the agreed price, the generator must meet its commitment to compensate the consumer. The consumer may be compensated by just receiving the price difference, which allows him to buy the electricity in the market at the agreed price, or by actually receiving the electricity "physically". As long as generators and consumers can rely on an underlying "physical" market, a purely financial contract carries the same value to them as a contract with physical commitments.

These financial contracts are easier for market players to manage, without losing any value as a risk management instrument. Physical contracts, on the other hand, create significant challenges in market design. Trying to maintain physical attributes in contracts puts overall economic efficiency critically at stake. In many markets, physical contracts prompt generators to commit transmission capacity in order to fulfil the requirements of their contracts. It has been difficult to integrate these commitments efficiently into a set of market rules and trading arrangements that allow for sound, cost-reflective pricing. Often these contracts were established prior to liberalisation and, as a consequence, they may be a barrier to competition.

Electricity is inherently volatile. Therefore, efficient outcomes require continuous pricing of all factors, including the value of transmission capacity. If a contract is physical and commits a generator and a retail company to a specific time frame, it must be backed by guaranteed transmission capacity for the same period. For example, two market players may agree to transmit electricity from A to B at a fixed price for a year. If electricity is suddenly offered at a lower price in area B than in A - even for a short period of time - the efficient outcome would be to use transmission capacity in the opposite direction (*i.e.* to transmit power from B to A) during that period. If the contract physically reserves the transmission capacity, such a shift of flow would not be possible: transmission and generation assets would be used less efficiently as a consequence. If market power is also a factor in areas A and/or B, the dominating player may even have an incentive to block transmission capacity to maintain this dominating position. In these kinds of circumstances, physical contracts become a real threat to the functioning of the market. In the Nordic market, initial market rules allowed for contracts with "physical" attributes to a certain extent; these rules were made obsolete with the development of credible spot and balancing markets.

Price volatility is a perceived risk factor for both generators and consumers. A generator runs the risk that the price will be too low to cover all costs. A consumer runs the risk that, at times, the electricity price will be too high compared with the benefits electricity consumption creates. A market for contracts offers generators and consumers an opportunity to meet and create a mutual risk hedge. Both parties will probably even be prepared to pay a premium to avoid risk. In this sense, contracts offer protection both for generators and consumers.

Power Utilities Use Contracts to Manage Risks

In a vertically integrated and regulated electricity industry, risks are managed through decisions within the utility and all costs are passed on to consumers. Unbundling breaks up the value chain thereby bringing risks into the open - making them transparent. Competition effectively shifts business risks from consumers to those who actually make decisions at each point of the value chain. In this unbundled and competitive electricity industry, contracts become one of the instruments that market players will use to manage the business risks, just as they are used by many other sectors with homogenous products. A range of such instruments are now available, their development having been driven by the needs of market players, new IT tools and recent economic research. Many of the risk management tools and contracts previously applied in other markets for commodities, currencies, stocks and bonds, are now being introduced in liberalised electricity markets. Competition and cost-reflective pricing of electricity are pre-requisites for successful electricity market liberalisation. Building on that, it is equally important that the unbundled industry have access to instruments to manage the risks that have now become transparent. Development of an effective market for financial contracts is an integrated part of a well-functioning electricity market.

Financial contracts are necessary for market players to operate, but they are also important from the perspective of regulating competition. Economic research demonstrates that contracts committing market players over various future time-spans serve another important purpose: they effectively minimise the scope or incentive for market power abuse. In the final planning for the actual operation, when all information is on the table, a dominating market player may be able to analyse all options and possibilities and to manipulate prices in day-ahead or real-time trading. By holding back generation capacity or simply by bidding higher than marginal costs, the player can increase the market price. The player may lose market share temporarily but even this factor can be considered and integrated into revenue calculations. If the same dominating player also made commitments prior to day-ahead, these commitments must also be taken into account. At the same time, if the player made a previous commitment to deliver electricity at a certain price, there is no chance to benefit from momentarily manipulating that price to a higher level.

Longer duration contracts remove the incentive to abuse a dominating position – or at least make the manipulation more complex and less profitable. From that perspective, an effective market for financial contracts is a sign of faith amongst participants in the market place. However, such an effective market will develop only if there are many players and if they trust that the underlying market price is not being manipulated by large incumbents. Thus, the transition period while a liquid financial market is developing becomes critical. The competitiveness of the underlying "physical" market depends on the presence of long-term contracts that limit the scope for market power abuse and the development of a liquid financial market depends on a credible underlying "physical" market. At the same time, maintaining previous long-term contracts negotiated prior to liberalisation may be an instrument for incumbents to maintain their market power.

The California experience suggests that it may not be a good idea to cancel all contracts at the implementation of a reformed wholesale market. During the transition period, it may be important to accept old "physical" bi-lateral contracts (as was done in Norway) as long as they are not allowed to distort the basic pricing principle.

An effective market for financial contracts is characterised by many factors. Some of the most critical factors are: the presence of institutions for trading; access to products that meet the needs for risk management; and sufficient liquidity (a market is liquid when users are able to buy and sell products at necessary volumes without problems and without affecting the price significantly). A market player who decides to buy a contract always wants to know that there are many other players willing to buy the contract back at the market price. One measure of liquidity is that the distance between the "best offer" and the "best bid" for each product is not too large (*i.e.* there is a "small price spread"). In Nord Pool's market for financial contracts there are so-called "market makers" in many of the most important products. A market maker commits to always give bids and offers of a certain volume at prices within a certain spread. Nord Pool compensates the market maker through various preferential conditions, including reduced trading fees.

The amount of market turnover compared with the underlying product that is physically being delivered is another liquidity indicator (Table 2 shows liquidity in different markets and different market segments, divided according to the institutions involved). Real-time trade is offered by the system or market operator. Day-ahead trade may be offered by an official system or market operator but may also be offered by a private commercial exchange. The segment of the market for longer duration contracts is often offered in a so-called "over-the-counter" (OTC) market, in which brokers organise interaction between market players enabling them to exchange bi-lateral contracts. Real exchanges for contracts have also developed in several electricity markets. With the development of on-line trading platforms and other sophisticated electronic management tools, the distinction between OTC and exchange trade has become less obvious. Depending on the legal framework of each country, the most important differences may be the legal requirements for exchanges to fulfil minimum levels of transparency, market monitoring and binding trading rules. The largest difference is the management of counter-party risk.

Data about liquidity in the real-time segment (Table 2) indicates differences in market design rather than providing any meaningful information about differences in liquidity. In Australia, the NEM is a centralised dispatching mechanism and NEMMCO has a 100% market share by definition. In PJM, market players are allowed to trade on their own, including meeting their balancing requirements: the 35% share indicates the level that balances their operations through this market clearing mechanism. In the Nordic market, the 3% real-time market liquidity indicates the actual turnover of regulating power needed to balance the Nordic electricity system. This is similar with the

Table 2

Liquidity in various electricity markets: Turnover in different market segments as share of total consumption, 2004

	England and Wales	Australia NEM	PJM	Nordic	Germany – European Energy Exchange
Real time	5%	100%	35%	3%	NA
Day-ahead	NA		26%	43%	11%
Further ahead: Exchange	NA	13%[1]	24%[3]	151%[5]	29%
Further ahead: Over-the-counter	NA	125%[2]	58%[4]	309%[6]	34%[7]

1. d-cypha trade, 2. Australian Financial Market Association, 3. NYMEX, 4. ICE, 5. Nord Pool, 6. Nord Pool Clearing, 7. EEX Clearing

Source: d-cypha trade, AFMA, FERC, Nord Pool, EEX

turnover in the England and Wales balancing market. In Britain, Sweden and Finland, market players are allowed to balance generation and consumption internally which decreases liquidity. In Norway and Denmark, all regulation must go through the market to minimise the regulating requirement on a system-wide level.

Day-ahead markets are the main reference in PJM and in the Nordic market. In the latter case, they accounted for 25% of contracts over a long period of time (PJM reports similar figures). The day-ahead share recently increased in the Nordic market to the current 43%, mainly as a result of a change in Nord Pool's trading fees. Considering the key position of Germany in the European electricity system, development of the German electricity exchange (the European Energy Exchange or EEX), is very important for the development of electricity trade in Europe. EEX has been in operation since 2000, with a relatively slow but constant increase in traded volumes in the day-ahead market segment; spot trading in this segment now corresponds to 11% of the total German consumption. With EEX and Germany at the centre of European electricity trade, liquidity data from Germany may reflect this role more than actually being an indicator of the effectiveness of the German market. European electricity traders perceive the EEX day-ahead market as a "balancing market" for trades in continental Europe.

The real challenge lies in developing liquid markets for contracts with longer durations. Some of the first experiences in the British market, with so-called "contracts for differences", developed into a relatively liquid market under the Pool trading arrangements. In the Nordic market, there was a strong increase in the liquidity of Nord Pool and OTC from the mid-1990s to 2002, at which point turnover in the financial contract market corresponded to more than eight times the total Nordic consumption. Turnover, measured in MWh, more or less collapsed during the Nordic drought but, when measured in traded values, the decrease was not equally dramatic. Liquidity began picking up again in mid-2003; in the first half of 2005, total turnover almost corresponded to the 2002 levels. Norwegian market players undertook some 50% of the trade in Nord Pool financial contracts and Nord Pool clearing; another 15% to 20% was undertaken by non-Nordic market players. Liquidity is also increasing in PJM and the FERC reports very rapid growth in the OTC market since mid-2004.[57]

In the British market, only the trade in the real-time balancing segment is a formal market. It is up to market players and commercial market institutions to create liquid markets for contracts with a longer duration. The markets with longer durations that have been established are not fully transparent and seem to be fairly illiquid. The exchanges UKPX and the International Petroleum Exchange organise electricity trading in the British market. Liquidity in these markets and in OTC trading is assessed by commercial consultancies. Market players in the British market seem to prefer a much less dynamic and costly risk management strategy of integrating generation and retail businesses through mergers and acquisitions. In such a situation, liquidity in terms of the possibility to trade real physical assets becomes even more crucial. This part of the British market is liquid, and indeed the competitiveness of the final outcome in terms of prices is another important indicator of liquidity.

Another pre-requisite for creating liquidity is that contract specifications can be aligned in standardised products. A long-term contract could be a very thick document that requires substantial legal input but has the advantage of being designed to meet the exact needs of all parties. Liquidity requires that contracts are standardised in relatively few forms, depending on the needs of market players. A retail company has short-term volume risks (due to the uncertainty of customer consumption) and longer term volume risks (due to the risk of losing customers). A retail company faces the risk of prices that deviate from the sales contracts to final customers. On the shorter term, generators have price/volume risks related to the difference between the price of fuel and the other costs associated to operating the plant; on a longer term, they face the risk that the market price will not sufficiently compensate for investment cost and profit. For a hydro plant, there is a fuel-related risk tied to reservoir development.

57. *FERC, 2005.*

To fulfil these needs, a variety of contracts are on offer in all markets: daily, weekly, monthly, quarterly and annual contracts. In many markets, products are also divided into "peak" and "base" products, a base product being one for all hours of the contracted period and a peak product applying only during working hours on work days. The Nordic market is, so far, an important exception to this pattern; no standardised peak-load contracts are traded at this point in time. Due to the central role of hydro power in the Nordic market, there has not been great variation in the hourly prices - although this seems to be changing now.

These basic contracts for financial delivery during a specific period in the future are known as "forwards" or "futures". Forward contracts are generally traded OTC, with settlement at the time of delivery. Futures contracts are generally traded on an exchange, with continuous settlement of differences between the agreed price and the reference market price. In addition to these basic contracts, "swaps" may be offered - contracts that link electricity trade with other risk elements such as fuel prices.

Also, the notion of "value of optionality" is becoming more and more recognised in many risky business environments, including electricity markets. There is growing appreciation for the notion that there is a value in being able to postpone a decisive choice of action; thus, the option of postponing an action has a value. It is valuable for a trader who owns a futures contract to be able to postpone the decision to sell the contract until he knows how the price will develop. It is equally valuable for an investor to build a new generation plant in small steps, being able to wait with the next step until it is called for and financially a sound investment. For a consumer, it is valuable to have some time to consider a new contract offer. Markets for option contracts that standardise these risk elements - with respect to standard features, terms and conditions - are developing, but are still relatively illiquid.

Risks in the day-to-day management of electricity generation and consumption have led to the creation of liquidity for contracts extending two to three years into the future in some markets. Many commitments reach even further ahead in time. For example, an investor in a new production plant might prefer to hedge the risks of the investment in a contract that spans a substantial part of the plant's economic lifetime. In general, there are few buyers for such contracts, which indicates that the investors are not willing to accept the discounted price in such a contract - the risk premium that a buyer would require. The need for liquid financial markets in the context of new investments is not a matter of allowing investors to cover all risks; investment in power generation is inherently risky, as is the case for all other investments. What is important, however, is that an investor knows that there is a liquid market for sale of the power during actual operation and that there is also a liquid market that makes it possible to manage the risks of buying fuels and operating the plant.

Markets for financial electricity contracts took a serious blow in late 2001, in the wake of the collapse of the global electricity utility and trader, ENRON. ENRON was a very large player in many markets; in that sense it contributed substantially to the liquidity in many electricity markets. ENRON operated its own on-line trading platform on which it posted its own bids and offers across several markets and market segments. The disappearance of such a large market player decreased liquidity in its own right. However, the most important effect was the reinforced focus on another important risk parameter: namely, the risk that a trading partner cannot fulfil its commitments. Apart from volume and price risks, this is the third most important risk to consider. Counterparty risk, as it is known, attracted more attention after the ENRON collapse.

The standardised service offered to manage this risk is called "clearing", in which an institution offers to step in between two market players who have entered into a contract. The clearing institution manages the counter party risk of each market player by assessing credit-worthiness and collecting collateral. Its presence allows individual players to forego costly management of the credit-worthiness of all trading partners and to concentrate on fulfilling its own requirements from the clearing institution. The clearing institution holds a key role in market transactions: a default on a contract becomes the problem of the clearing institution rather than of the trading partner.

One of the differences between an exchange and an OTC market is that the exchange implicitly offers clearing, although many electricity markets also have some market institutions that offer clearing services for OTC traded contracts. Nord Pool offered clearing services for OTC contracts from 1997; ICE, the large and quickly growing OTC trading platform in the PJM market, also offers clearing services. Liquidity in the Nordic market did not suffer significantly from the ENRON collapse, even though ENRON was also an important player in this market. Contract clearing may have been a key factor in minimising the impact.

As liberalisation made risks more transparent – and an integrated part of the decision-making process – it also prompted the development of very sophisticated risk management instruments in some markets. One of the most important remaining challenges is that of identifying the drivers for liquid markets for financial contracts and enabling this development in all liberalised markets.

What are the main drivers behind the relatively liquid financial market in the Nordic countries? Many years of development lie behind the success of the liberalised market in Nordic countries. Norway liberalised its electricity market in 1991 but financial trading only really began taking off in 1998-2000. The ownership structure of Nord Pool may also have an influence on the development. Nord Pool Holding owns Nord Pools financial trading activities. In turn, Nord Pool Holding is owned by the Norwegian and Swedish TSOs, Statnett and Svenska Kraftnät. Even if OTC trading still has the largest share of the Nordic financial market, Nord Pool undoubtedly played a very important role in developing the relatively liquid Nordic market. However, commercially, Nord Pool has not been profitable. The relative success of the Nordic financial market could be partly due to the fact that it is backed by a certain level of support from the two state-owned TSOs. The character of risks associated with the large concentration of hydro power may also have an influence.

Liquid financial markets contribute an important share of the potential benefits of liberalisation. The sophisticated risk management instruments offered in liquid financial markets improve risk management within utilities while also creating a platform for developing new customer products. To the electricity consumer, electricity is by-and-large a standardised product: the only things that matter are quality, reliability, price and the level of service in connection with billing. Retail prices vary with the wholesale price of electricity and with the revenue the retailer requires to cover costs and make a profit. Billing is a relatively straight-forward activity with substantial economies of scale, so it is difficult to diversify electricity as a product in that dimension. The most obvious way to differentiate electricity as a product - and thereby add value to the consumer - is to offer risk management.

In a competitive retail market, consumer decisions on how to manage price risks are probably much more important for the actual annual price paid than the choice of retailer. The decision to sign a one- or two-year contract at a specific moment can lead to substantially different annual prices than a decision to accept the spot price. In this sense, a modern electricity retail company is more like an insurance company than anything else and the product for sale is risk management. Product development focuses on finding ways to offer customers the contracts that best fit their appetite for risk; this, in turn, requires a liquid market for financial contracts.

Contracts Offer Protection for Consumers

Electricity prices reflect the volatile nature of electricity, and will rise to very high levels from time to time. During periods in which the electricity system is constrained by available generation capacity to meet demand, the price will go up in a short spike. If the electricity system is constrained by energy supply (*e.g.* due to drought in a hydro-based system or high fuel prices), the price will increase to a higher level over a longer period of time. For large consumers with a high share of electricity costs in their overall costs, such price volatility is a real business risk. They may have committed to selling their products at prices that are not profitable or competitive if electricity prices soar beyond a certain level. On the other hand, prices can also fall to low levels - triggered by low fuel prices, abnormal temperatures or high inflow to hydro reservoirs. Taking advantage of periods with low prices would be the reward for taking the risk of not signing a contract with fixed prices over a longer time period (*e.g.* one or two years).

Small household consumers generally spend only a low share of total income on electricity. Compared with other risks they accept in dealing with loans for houses and cars, petrol prices, foreign currency, etc., the financial risk from electricity prices is often relatively low. On the other hand, the variations of real-time and day-ahead electricity prices may create uncertainty for household consumers who do not understand and closely follow the development of electricity markets.

A threat to the competitiveness of large industrial electricity consumers, coupled with uncertainty on the part of household consumers, can very quickly turn into strong political pressure to intervene in the market. There are, however, no price-risks that cannot be managed fully through contracts. The most normal contract is to agree on a fixed price for a fixed term. More sophisticated contracts exist, based on the spot price but with a cap (retailers are able to offer these contracts by using option contracts in the wholesale market). Many of the largest electricity consumers spread risk by contracting a certain share of their anticipated consumption in one form of contract and other shares in other forms. The key point is that contracts can effectively shield consumers from short-run price fluctuations

For example, a large industrial consumer may have established contracts to sell aluminium over the following two years. In a competitive retail market, the consumer can sign electricity contracts that fix electricity price at levels that render these deals profitable. If the electricity price turns out to be very high, the aluminium smelter may consider re-selling the electricity in the market and fulfilling its commitments to supply aluminium through the world market for aluminium. This happened to a great extent in Norway during the winter drought in 2002/2003.

To reduce risk associated with the volatility of electricity, a household consumer might consider whether it is worthwhile to pay an insurance premium to avoid fluctuations in monthly electricity bills. If not, the consumer should sign for a contract based on the spot price plus costs for billing. If the consumer prefers to avoid risk of fluctuating monthly bills, the best option may be to consider a one- or two-year contract at fixed price.

It is evident that contracts can offer consumers protection from the inherent price volatility of electricity: but what about the question of political intervention if prices change dramatically? If the underlying wholesale price is competitive and the retail market itself is competitive, a political intervention to protect consumers would primarily be a risk-management decision taken on their behalf, and not necessarily an efficient decision. On top of that, a political intervention carries a cost in the form of lost credibility. The market foundation is undermined, potentially to a large extent and for a long period – well after the political intervention, and its underlying causes, have ended. Political intervention also withdraws the added value that comes from freedom of choice. In wholesale markets, it is not a sustainable solution for consumer protection.

Other critical decisions are necessary in relation to consumer protection. As many consumers will not be very active in retail markets, at least not in an initial phase, it becomes important to define how these customers are treated, *i.e.* what type of contracts are consumers left with as default? An option used in some markets, including New South Wales, Denmark and Spain, has been to maintain regulated tariffs for consumers who do not switch from the incumbent supplier. This helps protect the consumer, but it makes it very difficult for retailers in the market to compete with these regulated tariffs. If the benchmark for regulation is set at a low level, it will be impossible for a competitive retail company to propose an alternative offer that is sufficiently below the regulated tariff. In the short term, this may seem to be in the consumer's interest. But, if the result is that competition is undermined – and thereby also begins undermining innovation and development – it may serve the consumer poorly in the longer run. A tariff that undercuts a competitive market price will probably also be economically unsustainable; eventually it would have to be based on cross-subsidisation from other activities such as an associated network business.

The issue of default contracts also arose during the Nordic drought in the winter of 2002/03. In Norway, most residential consumers traditionally purchase electricity under a contract form in which prices can change from week to week, depending on the wholesale price (this flexible contract is the standard for historical reasons). During the winter of 2002/03, these flexible contracts created a situation in which retail electricity prices increased at unprecedented rates, within weeks of the increased wholesale prices. Consumers responded by reducing consumption to a certain extent, but it also raised intense public debate about the functioning of the market and how to address the social consequences. Sweden faced a similar drought and saw more or less the same increases in wholesale prices. But in Sweden, the norm for residential customers is a one-year contract at a fixed price. As a consequence, retail electricity prices increased only following a substantial time-lag and only at a fraction of the short-duration increases experienced in Norway. The ensuing public debate in Sweden had a fundamentally different character and did not include social aspects. Unsurprisingly, residential customers in Sweden barely responded to the price by reducing consumption. In the wake of the drought (during the 3rd and 4th quarters of 2004), the share of Norwegian residential consumers with variable price contracts decreased from almost 95% to 85%. However, it has since increased again, indicating that Norwegian households still prefer variable price contracts.[58]

In most liberalised markets, the approach is to let all retail prices be determined through competition within the retail market itself. This raises concerns for consumers who do not want to take the time to evaluate differing electricity contracts. The transaction time and cost connected with actual switching also become critical factors. In some markets a centralised dissemination of price information eases the process by allowing consumers to compare retail prices; other markets use other approaches, but always with the same goal of facilitating comparison – and often through on-line services. The United Kingdom Utilities Act 2000 established an independent body, Energywatch, to protect and promote consumer interests by providing impartial information and advice, and by taking up complaints from consumers. The Norwegian Competition Authority publishes detailed price information on its Web site, as does the Swedish Consumer Agency. In Denmark, the electricity association collects and publishes retail price information. In Finland, incumbent retailers in a specific local area are required by law to list competitor prices. Various public utility commissions in the United States, in those states where there is freedom of choice, operate Web sites with price information (*e.g.* the Public Utility Commission of Pennsylvania).

Consumers will only be protected through contracts if the wholesale market behind such contracts is competitive – and if competition between retail companies leads to product offers that meet consumer needs. In that sense, consumers are best protected through a well-functioning, wholesale market with cost-reflective pricing, liquid financial markets and competitive retail markets. Several markets have found the recipe for relatively successful wholesale markets. Experiences with competitive retail markets are more uncertain.

---

58. NVE, 2005.

Retail Competition

A competitive retail market is the last link to ensure that consumers can receive some of the benefit derived along the value chain in electricity market liberalisation. In fact, effective retail competition creates a natural protection for consumers. Retail companies under pressure from competition may develop innovative contracts and products that create even more added value from liberalisation. As much as it is a crucial part of the liberalisation package, retail competition has proven to be one of the more difficult challenges and it carries the cost of rolling out the necessary infrastructure to make retail competition feasible. Delivering benefits from competition to small residential consumers is the most difficult as the costs relative to the benefits may be significant. Installing interval meters that can be read remotely on a daily basis is the option that adds the most value, but it is relatively costly to execute and operate. The alternative for small consumers is a system that uses calculated load profiles based on monthly, quarterly or annual meter readings. But this is also costly to manage. Moreover, it creates distorted incentives and it fails to enable a pass-through of hourly prices.

In most jurisdictions in the British, Nordic, PJM and Australian markets, all consumers have the right to change retail supplier (Figure 9).

Figure 9

Year in which all consumers are allowed to switch retailer

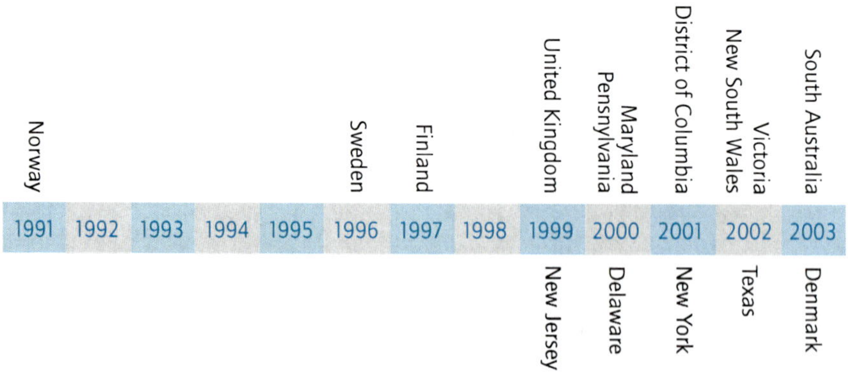

Examining the shares of consumers who have exercised this right shows interesting trends, even though the switching rates are not directly comparable (Table 3). Data is collected from various sources and the oldest dates back to 2001 in the Finnish case.

Table 3

Switching rates amongst contestable customers: Number of customers no longer supplied by incumbent retailer as share of total number of residential and non-residential customers

	Norway	Sweden	Finland	England & Wales	New Jersey	Pennsylvania
Customer class	*By 4th quarter 2004*	*By 2004*	*During 2001*	*By March 2003*	*By 2004*	*By 2004*
Residential	24%	32%	1%	38%	18%	6%
Non-residential	31%	N.A.	3%	N.A.		15%

	Maryland	New York	D.C.	Texas	Australia - NEM	Denmark
Customer class	*By 2004*	*By 2004*	*By 2004*	*By 2004*	*By May 2005*	*During 1st quarter 2003*
Residential	4%	5%	11%	10%	16%	1%
Non-residential	29%	33%	38%	62%		8%

Source: NEMMCO, OFGEM, Elfor, NVE, STEM, Finergy, Pfeifenberger, Wharton & Schumacher (2004)

Even though these data are not directly comparable, the tables illustrate several general tendencies. Non-residential consumption generally represents between 50% and 75% of total electricity consumption in a given country. In general, non-residential consumers with higher consumption have substantially higher switching rates; in fact, their switching rates can be high enough to undermine the market share of a particular retailer in this segment. According to the fourth benchmarking report carried out by the European Commission in 2005, more than 50% of large industrial consumers switched supplier since they acquired the right to do so in the following countries: United Kingdom, Norway, Sweden, Finland and Denmark.

In contrast, residential switching rates are relatively low, except in countries with many years of full retail contestability, such as the United Kingdom, Norway and Sweden. It must be noted that these significant residential switching rates can also be attributed to other factors. In the United Kingdom, electricity and gas retail have been pooled, which makes switching more attractive for many consumers. In Norway and Sweden, residential electricity consumption is high due to the high shares of electric heating.

A consumer's decision to switch supplier will typically depend on how much he can gain and how much it costs to switch. The benefits will depend on the level of competition between retail companies. Retail companies under pressure from competition will cut profits to a minimum, optimise customer management and develop new innovative products, probably focusing on risk management. The decision to switch typically depends on the direct switching costs and on the efforts necessary to assess offers and perform the actual switching process. In a policy context, the most straightforward path to generating the necessary competition in the retail sector has been by minimising the costs of switching and lowering the cost of new entry through the creation of a framework for an efficient balancing market and a liquid financial market.

A first serious barrier to switching is the actual switching management. A local network company with its own retail arm will not have incentives to make consumer switching easy. In most markets, setting up standardised procedures that correspond with regulated third-party access on the wholesale level has been an integral part of the regulatory framework. In Australia, NEMMCO manages a fully centralised system for consumer switching.

Timely billing is a recurring problem in many markets: there are typically long delays in transmitting correct billing data from local network operators to commercial retail companies. This is a constant threat to truly independent retail companies. Free flow of information is another important barrier. Local network companies usually possess all historical metering data; unless competing retailers have easy access to the same data, an incumbent retail supplier may stand to benefit from this information.

Another important issue, particularly for residential and small consumers, is the necessary metering equipment. Electricity consumption priced according to the normal hourly or half-hourly market settlement requires metering at the same time intervals, and the metering data must be passed on for processing within the normal settlement cycle, which is usually a daily cycle. At larger consumers, this function is carried out by remotely read interval meters installed on site (in the Nordic market, more than 50% of the load is metered by remotely read interval meters).[59] The price of installing, operating and maintaining this equipment is easily justified for higher levels of consumption. For low commercial and residential consumption, the cost-benefit comparison is more questionable.

The alternative attempted is to construct a consumption profile for small consumers, which can be used to compute electricity consumption by a specific residential consumer in a specific hour according to a standard formula. The

59. *Nordel, 2004b.*

computation will always be an approximation, however. Increasing the level of sophistication would improve the approximation, but it would also increase the costs of computing and managing the system. This approach, known as load profiling, prompted high shares of residential retail switching in England and Wales, Norway and Sweden. However, this practice will inevitably introduce distorting incentives - at least to a certain extent. Some particular consumer profiles will benefit at the cost of other types of profiles. For example, consumers with a load profile which has higher peaks than the assumed load profile - and with higher consumption during high price hours - will benefit. An example of marketing by retail companies that illustrates some of the consequences of such distortions comes from Australia, where retailers offered cash-back incentives to encourage customers to buy air-conditioners. A retail company may benefit from peaky demand in the market segment with load profiles. This, in turn, increases peak demand in the total NEM.

Recently, more and more jurisdictions are giving serious consideration to full-scale roll-out of remotely read interval meters. In Italy, the utility Enel decided to install remote metering equipment in approximately 30 million households, at an estimated cost of EUR 2.1 billion (corresponding to a unit price of EUR 70).[60] By the end of 2004, more than 20 million meters had been installed.[61] Several local network companies in the Nordic region initiated similar full-scale, roll-out projects. The local Danish network company, Sydvest Energi, aims to install remote metering for all of its more than 150.000 consumers by the end of 2007.[62] The main motivation for these projects is that they lower the administrative costs of managing the metering process while also contributing to improved function of the electricity market. Experiments and research on intelligent meters and communication via the electricity grid are ongoing in many countries.

In the European Union, the Electricity Market Directive commits all member countries to provide all consumers with free choice of retail supplier by 1 January 2007. But many countries still debate the overall costs and benefits of full consumer contestability. In 2001, the Government of Queensland released a report assessing the costs and benefits of full retail competition and subsequently decided not to introduce the concept. It should be noted that the only benefits regarded in the report were those that resulted directly from the assessed savings due to competition; more dynamic effects, such as those from product innovation, were not taken into account. Full retail contestability was also abandoned in several U.S. states.

---

60. *IEA, 2003b.*

61. *ENEL, 2005.*

62. *Sydvest Energy, 2005.*

Recognising that consumers without the freedom to choose are essentially captive to electricity suppliers, several alternative approaches have been suggested to address this issue. One of the most serious threats of leaving residential customers in a permanently captive state is that they will become subject to cross-subsidisation to large consumers. One approach is to allow competitive bidding for the right to supply captive customers, but it carries the risk that the incumbent utility will have an important information advantage. The debate on full retail competition remains inconclusive, but more or less all consumers in the relatively successful markets (the United Kingdom, Australia, the Nordic market, and PJM) have the freedom to choose, and many are using this freedom. In the end, it is a matter of considering two inter-related elements: a) how best to protect the consumer from inefficiency and cross-subsidisation; and b) how best prepare for innovation opportunities that future technology development may offer.

INVESTMENT IN GENERATION AND TRANSMISSION

Liberalised electricity markets are improving the efficiency of electricity system operations in many different parts of the value chain, from generation to consumption. Competition and cost-reflective pricing keeps pressure on all decisions and provides a clear signal for actions. However, operation costs represent only a share of total electricity costs. A substantial share of the electricity bill paid by the electricity consumer comes from financing generation and transmission assets. Improving decision-making processes for investment related to generation capacity may be one of the most significant benefits of liberalisation. But it is also one of the most serious challenges in liberalised electricity markets. Many investment projects have long lead times and carry financial implications for several decades after implementation. In a transitional phase while markets are established, there will be uncertainty that may undermine the investment climate. A liberalisation process may end in a "Catch-22" situation in which successful liberalisation depends on investments and investments depend on a successful transition to liberalised markets. Investments in power generation seem to be the big test for the development of robust and sustainable markets.

Many markets, including the Nordic, benefited from an initial overcapacity, which initially reduces pressure on the market, allowing it to pass through a transitional phase while market institutions are developed. Policy makers have time to build credibility for the market without being pressured by a tight supply-demand balance. At the same time, overcapacity also stresses the need for improvement. Markets with overcapacity are also often characterised by relatively slow demand growth, which implies that it is more difficult to identify the right timing for new investment and when tightness kicks in after some years, it will still lead to political pressure.

In contrast, markets that start out with a tight situation, such as South Australia and Ontario, are brought into a particularly challenging situation shortly after market launch. In essence, it is put to the test before it can develop properly. But the high growth in demand that caused the tight balance, as in South Australia, creates a situation that allows for swift and clear price signals for investment.

All in all, it is inevitable that new investments will be needed at some point, and that point will be a test to liberalised markets, regardless of the point of departure.

Market-driven Investment in Generation

A competitive bidding process determines prices in liberalised markets. Consumers and generators find the marginal resource that balances demand and supply, and a price is established. In periods with low demand, prices will be set by the generation plants with the lowest marginal costs. In periods with higher and peak loads, the generation plants with the higher marginal costs set the price. During the few hours with very high demand, the most expensive resources set the price - usually plants with low capital costs and high marginal costs, such as open cycle gas turbines (OCGT). Price can also be set by a reduction or shift of demand if consumption has a lower value to some consumers than the alternative market price. Marginal prices are set where marginal costs equal marginal benefits.

A generator under competition will be willing to produce at a price that pays the cost of each additional MWh, but this marginal cost will not cover depreciation or provide returns on the invested capital. Generation plants recover invested capital during periods in which the price is set by the more expensive plants. Thus, plants with high capital costs must operate most hours of the year to be profitable, even if marginal costs are low. These base-load plants will recover the invested capital when prices are set by plants with higher marginal cost. Mid-merit plants will only recover the invested capital during hours in which the most expensive resources are setting the price. The most expensive resources are only activated during the very few situations with very tight supply/demand balance.

Traditionally, resources for peak load rely on generation plants such as OCGTs. Thus, the profitability of investment in OCGTs depends on the owner's ability to bid OCGTs into the market at prices above marginal costs. This is not usually a problem because the owner of this "last" resort will also have substantial market power in the specific hours within which it is needed. The owner will be able to collect a scarcity rent. On the other hand, this market power may pose a threat to the economic efficiency of the entire market and raise political considerations. An important point for the functioning of the market is that a generator should never be the last resort.

The value of electricity consumption is not indefinite. Electricity is used for millions of different purposes by millions of different users. Each user will assign a different value to each purpose. It ranges from the value of being able to heat a private swimming pool on a Tuesday at noon to being able to safely perform heart surgery. At a certain price, there will be some consumer who is willing to make the effort to reduce consumption or shift it to another point in time. If market rules are adjusted to make it as easy as possible for the consumer to participate, more consumers are likely to be able to respond and lower price will result. Initially in liberalised markets, large industrial consumers are the most relevant, depending on their marginal benefit of consumption and on the costs of managing load reduction in terms of equipment, organisation and time. Market rules that allow for smooth participation of the demand side will effectively "cap" the market power of the generator with the last pivotal supply.

If cost-reflective pricing in competitive markets can effectively incorporate a portion of the demand side in system balancing, liberalisation has, in effect, created a new resource. This illustrates well the new investment paradigm that is developing in liberalised markets. Technologies are seen in new roles; the previous division into base-, mid-, and peak-load technologies is less clear. Risks have become more transparent and are shifted from consumers to those market players who make decisions on investment and operation. As a result, capital-intensive technologies with long construction times are viewed with much more scepticism - even if marginal costs are low. On the other hand, cross-border trade is increasing and now provides a larger market for capital-intensive technologies. Access to larger, growing markets may sufficiently reduce the risks of large, up-front investments.

Still, various barriers and complexities stall the development of markets that are robust enough to balance supply and demand in extremely tight situations. First of all, there is a strong, traditional focus on the supply side in the electricity sector - partly because it is costly and complicated to involve consumers in decision processes. This focus on supply also affected the establishment of market rules: they are rarely conceived with careful considerations on enabling demand-side participation. The policy perspective reveals an inherent transitional problem. To be able to manage their own active participation in the market, consumers will need to invest in equipment, education and time. Most consumers will also have a marginal benefit of consumption that is substantially above the market price during normal conditions. Average market prices are often in the range of EUR 20 /MWh to EUR 40 /MWh: but consumers with the lowest marginal benefits may only be willing to shift demand at prices of 10 or 100 times the average price. This implies that consumers cannot be expected to respond to prices before they observe some price spikes in the market. Ergo, initially, some generators may be put in a position in which they are the last resort and have full market power. From a political point of view, this may appear unacceptable - but intervention may, on the other hand, undermine the framework upon which active demand participation must be built. This is illustrated by the impact of price caps, set at too low of level, on investments in peak generation plants. In that sense, political intervention motivated by fear of a market failure becomes self-fulfilling.

The severity of the problem may be even greater if there is risk of real shortage during this transitional phase. If the system is faced with several sudden rare events, it may be necessary to shed load by force to keep the rest of the system operational. It is, however, not socially acceptable to regard forced outages as a natural way to balance supply and demand in electricity markets.

As a consequence, many markets introduced price caps. This limits a dominating player's room to manoeuvre, but it also effectively restricts overall ability to find the most efficient balances between demand and supply. If the price cap is low, it also sends a clear signal to the market that there is a will to intervene. This may have an even more discouraging effect than the price cap itself, reducing willingness to invest in generation, especially in peaking plants, and in demand participation.

But restricting market power is not the only reason for introducing price caps. If forced load shedding is the final resource, then this should at least be given a value. The problem originates from the fact that consumers do not disclose the values they see from consuming electricity in the first place. So the value or costs of forced load shedding is, by definition, unknown. This so-called "value of lost load" (VOLL) needs to be assessed. Some countries actively use VOLL assessments to calculate price caps and to determine the benefits of other initiatives to improve system security. Table 4 shows the different price caps used in PJM, Australian, British and Nordic markets.

Table 4

Price caps in PJM, Australian, British and Nordic markets /MWh

PJM	Australia	Britain	Denmark	Finland	Norway	Sweden
USD 1 000	AUD 10 000	None	None	None	NOK 50 000	SEK 20 000
EUR 830	EUR 6 250				EUR 6 300	EUR 2 100

PJM has a price cap that is mainly designed to act as a shield against market power abuse. Most liberalised markets in United States have price caps of the same magnitude. In Australia, Sweden and Norway, the price caps are motivated by considerations about VOLL. In Australia, the price cap was increased from AUD 5 000 to AUD 10 000 in 2001, when it turned out that the lower cap was used too often. In the Nordic market the spot exchange,

Nord Pool, also has a limit in its system price of NOK 10 000 /MWh (app. EUR 1250 /MWh); however Nord Pool regards this not as a price cap but as a technical limit, which will be increased immediately if need be.

In Sweden, there seemed to be a real risk that supply could not meet demand without forced outages, particularly after a few very tight situations in the winters of 2000 and 2001. One of the root causes of the problem was that some older, oil-fired plants had been mothballed. The Swedish government commissioned the TSO, Svenska Kraftnät, to conduct a study on investment in new capacity in the open market and propose solutions. The solution suggested - and accepted by the government - was to give Svenska Kraftnät the authority to purchase additional reserve capacity over a transitional period from 2003 to 2008. The rules for activating the reserves are carefully designed to minimise the distortive impact on the market. For example, if the capacity is activated, it is at prices increasing from SEK 8 000 /MWh to SEK 15 000 /MWh over the five year period. The purchase of reserves takes place in a competitive tendering process and an effort is made to attract as many resources from the demand side as possible to participate in the bid.[63] It should be noted that Svenska Kraftnät's study did not include any discussion of their ownership of gas turbines for operational reserves and the role this might play in creating transparency and predictability for investors.

In the Australian, British and Nordic markets, governments clearly stated their intent that market price signals will provide market players with the right incentives for investment. Except for the transitional arrangement in Sweden, no additional financial incentives will be used. These markets are "one-price-only" or "energy-only" markets. It should be noted that it is false to think that capacity will not receive remuneration in energy-only markets. A one-price-only market is intended to provide remuneration for both variable and fixed costs through the same price, which corresponds to the normal pricing principles for most other products.

In the Netherlands, the government initiated a study on security of supply, including whether one-price-only markets create incentives for adequate generation capacity. An independent governmental agency that carries out economic policy analysis, the CPB (the Netherlands Bureau for Economic Policy Analysis) conducted a cost-benefit analysis that concluded that the costs of a specific capacity measure could not be justified by the benefits.[64] Instead, the Dutch government focused on creating a regulatory and market framework that provides transparency and clear price signals. It also

63. *Svenska Kraftnät, 2002.*

64. *CPB, 2004.*

authorised the TSO, TenneT, to establish a market for last resort capacity reserves along the same lines as the transitional arrangement in Sweden.[65]

Production capacity is now being built in the Australian, Nordic and British markets. The South Australian market was put to a test quickly after its launch, when prices spiked to reflect the scarcity of supply. The rising prices raised political concerns, but the only intervention was to double the price cap. Investments were made in peak load plants, which was also the appropriate technology to meet the problem of a rapid increase in peak demand. The Nordic market has experienced a long period of overcapacity and no new investment, followed by a period of tighter supply/demand balance, during which investors hesitated to make investment decisions. The only substantial new investments were in subsidised wind turbines, mainly in Denmark. Today, several new plants are under construction: a new 1 600 MW nuclear unit is under way in Finland; Sweden is building a new 400 MW combined heat and power (CHP) plant; and Norway decided to build two new combined cycle gas turbines.

In many countries, much political concern remains for the barriers to timely and efficient investment, including whether a one-price-only market will give sufficient remuneration for the invested capital. These concerns are mirrored in Australia, the United Kingdom and the Nordic countries, where the situations are monitored carefully. However, there are – as yet – still no failures to point to. Investments in liberalised markets are made "just in time" *i.e.* when there is a real need. So far, there is no reason to believe that cost-reflective pricing, real competition and good market institutions will fail to send signals for timely and efficient investments. Instead of creating extra incentives to improve investment decisions through intervention, policy makers can adopt other and more obvious roles to remove barriers to investment.

In the Nordic market, environmental policy caused substantial uncertainty for investors. In Sweden and Norway, respectively, long debates have ensued about nuclear phase-out and the use of natural gas for power generation. In Denmark, subsidies for renewable technologies led to a massive increase in renewable generation capacity, including 2 500 MW of intermittent wind capacity. The Nordic market has, so far, managed the situation despite such levels of uncertainty. Apart from the political reasons for hindering use of certain technologies and supporting others, there is also the even more important problem of acquiring permission to build. The Not-In-My-Back-Yard (NIMBY) and the Build-Absolutely-Nothing-Anywhere-Near-Anybody (BANANA) syndromes are an ever-present part of the investment decision

65. *Boot, 2005.*

process in the electricity sector. Smooth and transparent approval procedures for new generation plants and transmission lines are central to creating a framework that allows for adequate, timely and efficient investment.

Box 4 . Market-driven investment in Australia

The Australian National Electricity Market (NEM) started operating in December 1998. Average monthly prices from 1999 to 2004 are shown in Figure 10. Peak demand in Australia is during the summer - due to high use of air conditioning - and is increasing at a rapid rate. During the first years after the market opening, the balance between supply and demand was relatively tight. This is reflected well in the prices, with relatively high average prices particularly during the summer months and particularly in South Australia in the summer 2000/01.

The NEM is based on a one-price-only principle in which the electricity pricing determined by normal trading arrangements provides remuneration both for variable and fixed costs. Timing, volume,

Figure 10

Average monthly prices in national electricity market, Australia

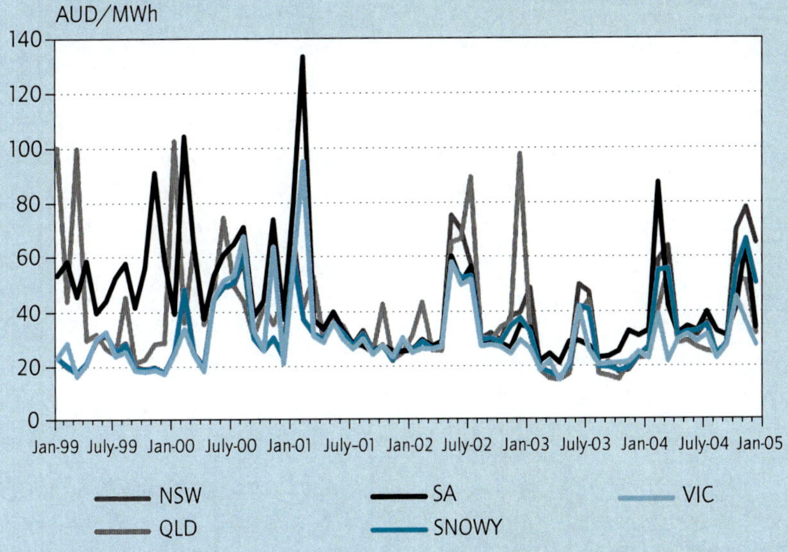

Source: NEMMCO

technology and location are left to investors to decide. Figure 11 shows the development of installed generation capacity at principal power stations in Australia - past, present and predicted.

Figure 11

Installed capacity principal power stations - 30 June, MW

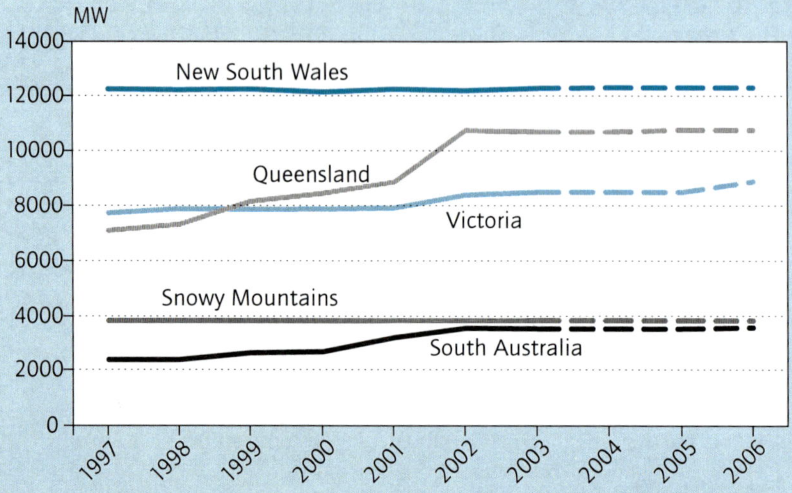

Source: ESAA and NEMMCO, predicted number from NEMMCO Statement of Opportunity, 2004.

Generation capacity in South Australia increased by 49% from 1998 to 2003 adding 1 133 MW to the total installed principal generation capacity. Almost half of that capacity was in the form of OCGTs for peaking purposes. The rapid increase in peak demand in South Australia, e.g. due to increased use of air-conditioning, created a tight balance between supply and demand during peak load periods, particularly in the summers from 1998 to 2001. This tightness was reflected in NEM prices, which peaked up to the price cap at AUD 5 000 /MWh in several hours during the first years of the new market (Figure 12). The price cap was then increased to AUD 10 000 /MWh.

South Australia was particularly in need of more generation capacity suitable for peaking purposes. OCGTs offer three advantages that make them well suited for this purpose: they have very low investment costs per kW installed capacity; they can be extended in small incremental steps;

Figure 12

Price duration curves for the highest percent in South Australia

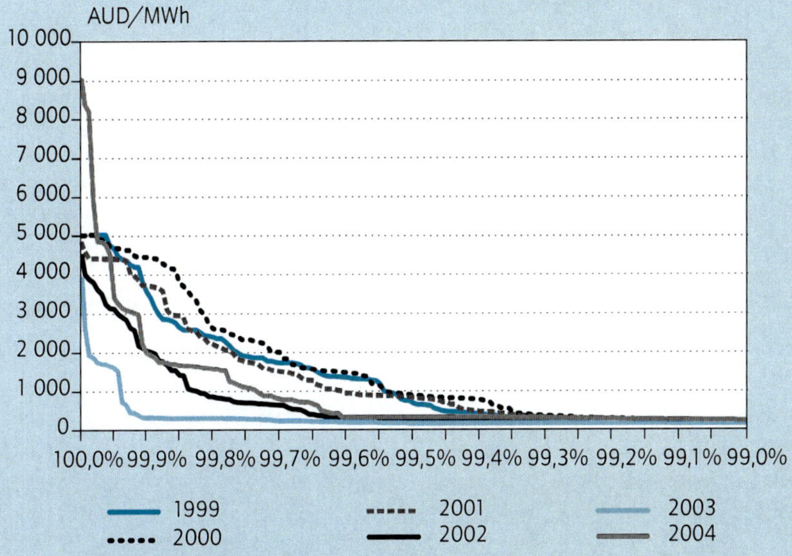

Source: NEMMCO

and they are very easy and fast to build compared to other conventional technologies. The relatively high cost of OCGT operation is less important as they will only run on rare occasions. Key financial features of gas turbines in Australia are in the order of AUD 500 /kw installed capacity and short-run marginal costs of AUD 40 MWh.[66] Thus, a 100 MW gas turbine would cost AUD 50 million. In 2004, the price rose above AUD 40 /MW during 1 790 hours in South Australia. A 100 MW gas turbine would be expected to operate during those hours and would receive some AUD 29 million, after fuel costs have been paid. In 2003, the same plant would only be expected to operate during 317 hours, with resulting revenue of some AUD 7 million after fuel. This shows that it is, indeed, risky to invest in a generation plant only for peaking purposes – particularly when expected income varies so much from year to year. On the other hand, if the investor analysis of the need for additional peaking capacity is correct there is a pricing and trading arrangement in place that rewards the high risk with very profitable rates of returns. It would take only two years similar to 2004 to finance the entire investment.

66. IEA, 2003a.

Price spikes in the South Australian market raised political concerns and inquiries, prompting the establishment of a South Australian Government Taskforce, which published its report in June 2001. The report addressed issues of market power and inappropriate market rules and considered price caps and other direct market interventions. Regarding possibilities to intervene, the report reads as follows:

> Further, the Task Force concluded that actions to change existing contracts could seriously damage [South Australia] in terms of investor confidence, future reliability of electricity supply including new investment urgently needed in South Australia, and retail market competition.[67]

Thus, investor response to market prices effectively overcame the tightness in the first years of the market and the South Australian Government actively stated that it would not intervene directly – it had confidence in general market principles. The joint federal and state decision to increase the price cap to AUD 10 000 /MWh was probably also key to incentivising the necessary market response; it was a much-needed signal of confidence in the market and of the governments' broad commitment.

In Victoria, generation capacity increased by 8% or 615 MW from 1998 to 2003, primarily reflecting investor response to market prices. Victoria's situation was similar to that of South Australia with a tight balance during peak periods in the first years of the NEM. Prior to a shortfall of capacity, due to strikes at major generating facilities in January 2000, there was a cap mechanism in Victoria, the "Industrial Relations Force Majeure". This may have deferred investments in peak capacity up to this point in time.[68] *But in 2001 and 2002, some 552 MW of peaking gas turbines were added to the system in Victoria.*

NEMMCO issues a so-called Statement of Opportunities *(SOO) every year on 31 July, with a brief update on 31 January. The SOO is a critical instrument to disseminate information and analysis on the status of key fundamental factors in the market to all market players and potential investors. It applies a very systematic and comprehensive methodology to collect, analyse and report the data in the following areas: demand information and demand forecasts including peak demand; generation capabilities that account for real conditions (including consequences of increased temperatures of cooling water during summer); and status of the transmission grid. It calculates*

67. South Australia Government Electricity Taskforce, 2001, page 4.
68. IEA, 2003a.

reserve requirements, based on all the input on demand, supply and transmission and on acceptable reliability requirements. Finally the demand, peak demand, generation capabilities and reserve requirements are used to calculate the future point in time at which maintaining the status quo would lead to a reserve deficit. The SOO thereby specifies future asset needs in terms of time, volume and location. It is then up to market participants to respond to this analysis and choose the appropriate technology.

The SOO also reports on plans and ongoing work on new generation and transmission assets. When a new asset is regarded as committed by NEMMCO, it is included in the computation of future reserve deficits. Thus, investors can carefully follow expected development in the total asset base.

The SOO 2004 predicted a reserve deficit of 320 MW in the joint Victorian and South Australian markets, which are closely integrated, by the summer of 2006/07.[69] This area has seen little investment in new generation capacity since 2002, at which point a project was initiated to connect the NEM with Tasmania and its hydro resources. In July 2003, NEMMCO and the Tasmanian Government signed a Memorandum of Understanding to integrate Tasmania into the NEM through a new 600 MW HVDC transmission line, known as Basslink. For the South Australian and Victoria markets, this will have a similar impact as would a new generation plant; the MOU has probably deferred new investments. According to SOO 2004, the Basslink project is committed to be comissioned by summer 2005/06 and is thus already included in the forecast reserve deficit of 320 MW for 2006/07. In the SOO update from January 2005, NEMMCO reports a new OCGT project of 312 MW, which they regard as committed to be brought into commercial operation in Victoria in December 2005.[70] It will thereby compensate for the forecast reserve deficit. There seems to be a will among market players to respond to market needs and the SOO dissemination process plays an important role to this end.

Generation capacity in Queensland has also increased substantially, mainly in the form of coal-fired generation capacity, which is financially most suitable for base load operation. The decisions behind these investments predate the opening of the NEM. According to the NEMMCO Statement of Opportunities 2004, Queensland had significant capacity reserves for the 2004/05 summer peak – almost 1 000 MW more than the required level for reliability. A new 750 MW coal-fired plant is

69. NEMMCO, 2004.

70. NEMMCO, 2005.

regarded as committed to go into commercial operation in 2007. The SOO 2004 projects a small capacity shortage for the 2009/10 summer peak and increasing thereafter. At present, most of the major generation companies in Queensland are state owned, although there has been private sector investment past 1998.

In spite of prices having been high at times (in fact, the highest on average in the NEM during the summer of 2004), there has been very little investment in New South Wales. A full 97% of all principal generation capacity in New South Wales is state owned. The state government blocked a private company proposal to build a small coal-fired power station – on greenhouse policy grounds. This was the first time such policy action was taken in Australia.

Figure 13

Price duration curves - National Electricity Market 2004

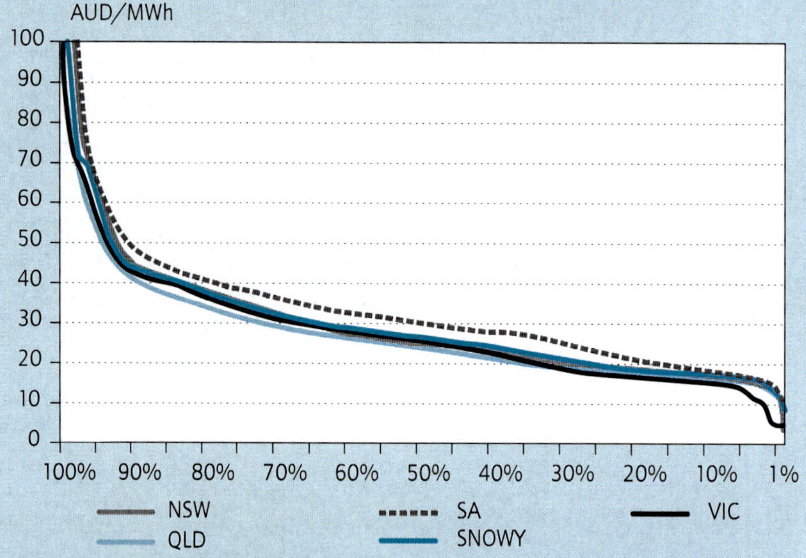

Source: NEMMCO

Figure 13 shows that even if prices are peaking at very high levels at times, prices remain below AUD 100 /MWh in 95% of the year's 17 520 half-hourly prices and below AUD 50 /MWh a full 90% of the time. In 2004, the annual average price across the entire NEM was AUD 38 /MWh. In any case, all market players – from generators to retailers – supplying household consumers are free to manage price variations through contracts to hedge risks. The final price in the NEM reflects the bid of the last dispatched marginal generator or consumer. Prices actually paid for large portions of the supply have been agreed through financial contracts between generators and retailers directly. However, the NEM price is the reference because all production and consumption will be settled at the NEM price, even though price fluctuations can be managed to the preference of the market players through contracts.

Box 5 . Market-driven investment in nuclear power in Finland

After almost a decade of careful planning, research and public debate, the Finnish generator Teollisuuden Voima Oy (TVO) started construction of its third nuclear unit at the Olkiluotot nuclear power plant in 2004 (the final decision to proceed was made at the end of 2003). The plant will have a capacity of 1 600 MW, generating 13.3 TWh per year with a 95% capacity factor. It is planned to go into commercial operation in mid-2009.

Finland is a part of the integrated Nordic market, which offers no specific remuneration for capacity in its one-price-only pricing principle. Thus, investments in the Nordic market will be financed through the contracts traded in the market place consisting of the day-ahead spot market, the exchange for financial contracts operated by Nord Pool and the bi-lateral over-the-counter (OTC) market. Nord Pool's day-ahead spot price is broadly regarded as the reference price in the Nordic market and is used as the reference in all other contracts with shorter or longer duration. The new Finnish nuclear unit will receive its entire remuneration from contracts in the market place and the Nord Pool spot price will be the reference.

The company and ownership structure of TVO make the financing of Finland's new nuclear reactor an interesting case concerning risk management and investment in nuclear power in competitive electricity markets. TVO is a part of the Pohjolan Voima Group, under the parent company Pohjolan Voima Oy (PVO), which owns some 60% of TVO shares. The rest are owned by the largest Finnish electric utility Fortum Oy (some 25%) and other smaller, municipally owned electric utilities. PVO was established by large industrial electricity consumers, mainly from the Finnish pulp and paper industry. These large industrial electricity consumers still represent the largest share of the owners but today PVO is also partly owned by local municipalities – either directly or through municipally owned utilities. Thus, large Finnish industrial electricity consumers have a very important stake in the new reactor. Even though public ownership – directly or indirectly – is still below majority, there is substantial public involvement in the nuclear project in that a large minority of PVO owners are municipalities, many of which also own important shares of TVO. Fortum, the second largest owner of TVO, is controlled by the Finnish state as a majority share owner.

The business of TVO is structured so that all electricity production passes from its nuclear power plants to its owners, according to owner shares at cost price.[71] *Ownership of TVO thus becomes a "physical hedge" towards fluctuations in electricity prices. The role and significance of such a physical hedge must by analysed in the context of other contracting opportunities offered in the Nordic market.*

In the Nordic market, all participants including TVO are free to make contracts of shorter and longer duration. That means that the "physical" delivery from TVO can be sold on to other market participants and that – if a price can be agreed – the risk of owning the physical assets in TVO can also be passed on to other market participants through financial contracts. The Nordic market is relatively liquid for financial contracts up to three years in advance. Within that framework, it is possible to undertake a quite sophisticated risk management strategy. Some contracts extend even further. In March 2005, one of the largest Swedish pulp and paper companies, Holmen AB, announced the signing of a 10-year contract with the Swedish utility Vattenfall. The contract commits Vattenfall to supply roughly half of the electricity demand from Holmen, some 1.5 TWh/year for the next 10 years. Holmen has its own generation capacity corresponding to roughly 1/3 of the consumption.[72]

The investment decision by the large industrial consumers owning large stakes in the new nuclear unit is probably an important part of their risk management strategies. As the Nordic market offers opportunities for advanced risk management, the linkage between Olkiluoto III and the electricity demand of the industrial owners should not be over stated, however. If Olkiluoto III ends up generating electricity at a long-run cost

71. TVO, 2005.
72. Holmen, 2005.

that is above the average spot price, the owners will incur a loss compared to the alternative. If the costs are below spot prices the investment will be profitable. Even if the investment is an important part of a risk management strategy, this does not change the fact that the investment decision will only be profitable if the analysis of future demand and supply in the Nordic market is accurate and correspondingly that the analysis of the plant costs is accurate.

Will the added capacity be needed and will the costs be covered by market prices? Nordel, the association of Nordic independent transmission system operators, publishes an annual report including basic supply and demand statistics and forecasts. In the annual report for 2003, Nordel reports that the forecast supply/demand balance for Finland in the winter 2007/08 is a deficit of production capacity of 795 MW.[73] *Nordel also projects power and energy balances three years ahead. In the projections for 2006, Nordel anticipates importing 8 TWh with normal precipitation conditions, 10 TWh with low precipitation and 18 TWh with an extremely dry year. At the same time, Nordel expects electricity consumption to increase at an average yearly rate of 1.2% until 2006.*[74] *All in all, there seems to be good reason to believe that there will be an internal Nordic demand for the extra 13 TWh that Olkiluoto III will bring to the market.*

The question, then, is if Olkiluoto III will be able to compete with alternative investments in the Nordic market and imports from Germany, Poland and Russia, for which interconnections are already in place, under construction (as is a new interconnection between Norway and the Netherlands) or in the planning stages (Finland and Estonia agreed to a new interconnection with PVO being one of the joint owners). In addition, a private company filed for permission by the Finnish Ministry of Trade and Industry to build a transmission cable between Finland and Russia. If approved, the cable will be able to feed 1 000 MW into the Finnish grid.

According to information provided for a joint IEA/NEA[75] *study on* Projected Costs of Generating Electricity - Update 2005, *the costs of a nuclear power plant in Finland is approximately EUR 40 /MWh.*[76] *These are the levelised lifetime costs of the plant, assuming a capacity factor of*

73. Nordel, 2004c.

74. Nordel, 2003b.

75. NEA is the Nuclear Energy Agency of the OECD.

76. IEA/NEA, 2005.

85%, 40 years lifetime and a 10% discount rate. TVO has operated the two reactors, Olkiluoto I & II, at average capacity factors of almost 96% from 2000 to 2004. According to the IEA/NEA study, the costs for the invested capital in nuclear power plants account for a large share of the overall costs. Assuming a 95% capacity factor for Olkiluoto III would almost decrease the costs by 10%.

The competitiveness of the plant will be tested against the electricity prices in Nord Pool. Figure 14 shows prices for forward contracts traded at Nord Pool for delivery three years ahead in time at the moment they are traded. Three years is the longest contract traded at Nord Pool and, thus, gives the best indication of the expected prices under normal conditions with normal levels in the hydro reservoirs. In 2002, the contract furthest ahead in time was the contract for 2005, in 2003 it was for 2006 etc.

Figure 14

Nord Pool forward prices, yearly contract, three years ahead

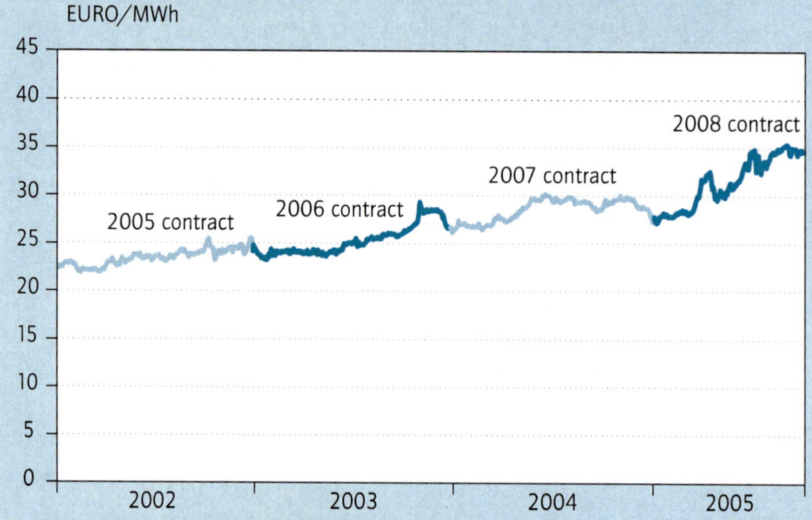

The three-year-forward prices have increased steadily since 2002, at a higher rate than can be explained by inflation. It seems to reflect the tighter supply/demand balance as forecasted by Nordel but also a whole range of other factors have an influence on the price. These include the increased costs of coal and gas and also the expected increased CO_2 emission costs.

Demand Participation as an Alternative

As long as some consumers participate actively in the balancing of supply and demand, a liberalised market can provide timely and adequate incentives for investments. On the other hand, lack of participation from consumers can undermine efficiency and reliability of electricity markets, ultimately raising prices for all users. Lack of demand participation remains one of the most serious weaknesses in the development of robust electricity markets. There are good reasons to be optimistic, however. Different types of consumption have different values for the consumer. Consumers are undoubtedly, in principle, willing to shift demand as a response to price; *i.e.* demand is price-elastic. The main problem or barrier is that managing the response is too complicated and costly: transaction costs are too high. If investment in equipment, organisation and time to manage the response cannot be offset by the revenues from responding, consumers can not be expected to be active. Figure 15 shows the importance and effects of active demand participation.

If some consumers can offset the transaction costs for some shares of their demand, they will become more responsive to prices; demand will become price-elastic, putting downward pressure on electricity prices. Demand participation will create or release resources that, in the end, can save investments in new generation plants. It can also curb market power of the generator with the last pivotal generation resources and add to overall system reliability. These factors are a major source of improved efficiency in competitive electricity markets.

In a time of transition, it is likely that - at least initially - only a small share of the demand side will be able and willing to respond to price. This will push the price a little lower than the inelastic clearing price. As the market matures, the amount of demand that finds it worthwhile to enable price response will increase. This development will stabilise at a level at which the alternative new generation plant is less expensive than the marginal benefit from consumption, taking revenues and transaction costs into account. The role of a price cap becomes critical in such a process. If the price cap is below the elastic clearing price in a mature market, it is unlikely that any demand participation will be realised. Even if the price cap is between the elastic and inelastic clearing prices, a price cap may undermine the development path towards more active demand participation.

Apart from the direct price effect to improve efficiency and market performance, demand participation also has important "volume effects". Firstly, it adds flexibility that will improve system reliability: the more demand resources that can be activated with the price instrument, the less likely it will

Figure 15

Demand participation improves market and system performance

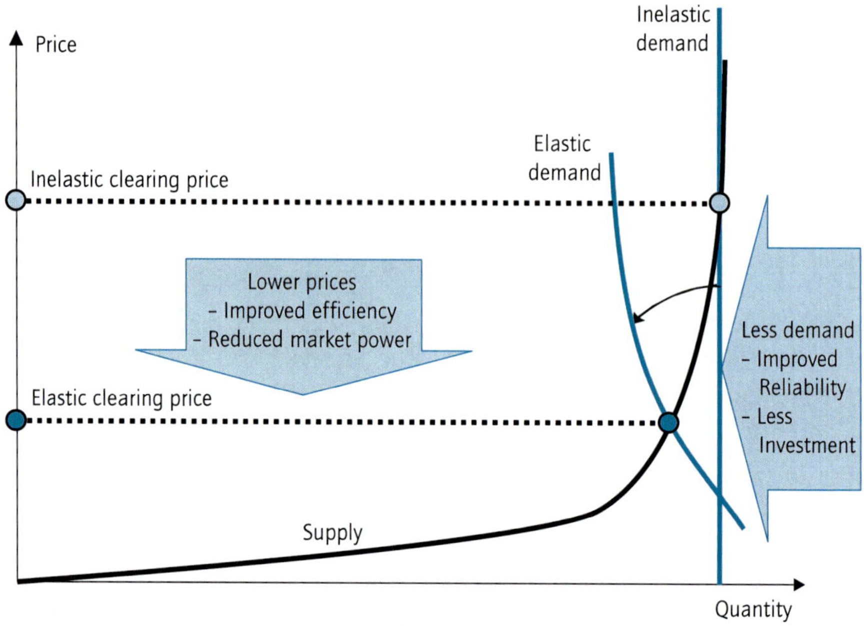

be that the system operator will be forced to non-voluntary shedding of load. It is also likely that clear price signals that reach the consumer will lead to real conservation of electricity and in any case more efficient use of energy. Investment in equipment, organisation and time that enable demand participation will, in most cases, improve the general level of energy management. Knowledge and control is a first prerequisite to efficient energy use, which will be in the financial interest of any consumer.

Transaction cost is the critical parameter for activating demand participation. Transaction costs come from upfront investment and day-to-day management. In general, transaction costs per MWh will decrease with the level of the consumption, *i.e.* transaction costs per MWh will be higher for small consumers and lower for large consumers. Initially, it is expected that demand participation will mainly be of interest to the largest industrial consumers. That being said, transaction costs are also highly dependent on the character of the consumption. Some industrial processes are highly automated, so no significant additional investment and management costs will be required. On the other hand, the marginal benefit of consumption may be relatively high. Some residential consumption, such as electric heating or air-conditioning, will require relatively high up-front investment costs to allow for full automation of demand participation. Once in place, however, the actual marginal benefit of consumption in a specific hour may be relatively low.

Reaction times and duration times also have a large impact on marginal benefits and transaction costs. Some industrial consumers may shift load at a relatively low loss of benefit, if it is with a day's notice, but the loss of benefit may be significantly higher if it is only with an hour or 15 minutes notice. On the other hand, a residential customer with electric heating and enabled to automatic load shifting is able to react on a very short notice without losing greater comfort than if it is with a day's notice. Different types of demand response will be best suited for different segments of the market. Some are best suited for day-ahead commitments, some are able to respond in real-time and others are willing to commit to a contract of longer duration to stand as operational reserve.

One of the operational challenges of active demand participation is that it is not possible to meter load shifting in the same way as electricity generation. Other methods must be considered to measure these "nega-WATTs". One approach is simply to regard resources from demand response to market prices as an additional statistical parameter to take into account when making demand forecasts. A retail company can offer a spot price product to its customers and then monitor the effect this has on the demand by its customers in total. Over time, the retail company will start to learn how to account for price when making demand forecasts, in the same way as it is today taking temperature and other similar factors into account. The retail company is then able to take this knowledge into account when submitting bids to the market.

Policy makers, regulators and system operators have important roles in making the access for demand participation as easy as possible, thereby minimising the transaction costs for consumers. Low price caps and political intervention in the market will deter demand participation. All market rules and trading arrangements should be considered with the demand side in mind, including design of products and services to meet TSOs need for reserves and other ancillary services. All TSOs in the Nordic, PJM, Australian and British markets allow consumers to bid into some of their competitive purchases of reserves and other ancillary services.

An interesting example of innovation in the wholesale market is a new bidding rule introduced by the Nordic power exchange (Nord Pool) to meet requests from consumers. Nord Pool now allows bids in the day-ahead spot market that specify a price and a volume, but not a time. This implies that the bidder commits to reduce load within any one hour in which the spot price is higher than the bid price. Thereby the consumer avoids having to give bids for several hours, with the risk of having to reduce load in several hours.

Retail companies are the link between the consumer and the wholesale market and, therefore, have a key role in making demand participation a reality. They will also be dependent on appropriate market rules and trading arrangements, but they will have to make the necessary product innovations that can link consumers to the wholesale market. Apart from the innovation triggered by competition, several countries have also considered how to contribute to this innovation process through R&D programmes and pilot projects. In PJM and many of its control areas, there are programmes to promote demand participation. In the Nordic region, TSOs are preparing action plans to support the development of demand participation.[77] The 2002 review of the Australian National Energy Market stresses the importance of demand response and suggests some action points.[78] A project on demand response resources under the IEA Implementing Agreement on Demand Side Management is exploring options and best practices of enabling demand participation.

There may also be a role for local network companies. Congestion points in distribution grids are not priced using normal locational pricing - not even in the nodal pricing system in PJM. A local network owner may consider giving consumers special incentives to shift load during peak conditions to avoid additional investment in the distribution grid. In some circumstances this can have a value that is significantly higher than the value reflected in the normal day-ahead and real-time pricing.

Table 5 shows demand participation that has been contractually committed by TSOs, observed in the market and assessed to be additionally available at a minimum. The numbers cover demand participation in many different forms, with many different duration times and many different reaction times.

It is difficult to observe demand participation with certainty. The point with liberalised markets is that decisions are made on a decentralised level. The demand side is free to participate in the market through bids. In Australia and in England & Wales, very low levels of demand participation have been observed with certainty: yet a participation corresponding to 1% of peak demand is an important contribution. It does, however, seem unlikely that all consumption in Australia has a marginal benefit corresponding to the most extreme price spikes during peak-load hours. There seems to be barriers to enable demand participation. The extreme peaks in Australia are driven by air-

77. Nordel, 2004a.

78. Council of Australian Governments Energy Market Review, 2002.

Table 5

Demand participation: Committed by system operators with minimum additional assessed and observed demand participation

	PJM	Nordic	England and Wales	Australia	Alberta (Canada)
Committed	3 598 MW	2 075 MW	4 329 MW*	NA	NA
Percentage of peak load	3%	3%	7%*		
Additional - observed and assessed	7 964 MW**	10 000 MW***	800 MW**	334 MW**	800 MW
Percentage of peak load	7%	15%	1%	1%	7%

** Britain ** Observed, *** Observed and assessed*
Source: PJM, OFGEM, NORDEL, NEMMCO.

conditioning. It seems unlikely that there are no large industrial consumers who would find it profitable to reduce some consumption during the most critical hours and save AUD 5 000 /MWh to AUD 10 000 /MWh. This lack of participation may be a sign of inadequate competition in the retail market. Another aspect that may contribute in Australia is the market design. There is a strong focus on real-time market operation, which provides a very flexible environment for supply but may make it more difficult for the demand side to participate. To many large industrial consumers that may be price responsive, the cost of responding on very short notice may be very high. In markets with day-ahead spot fixations, it is possible for consumers to lock in the "trade" one day in advance and they are, thereby, left some time to plan the reduction.

Data from the Nordic market are drawn from a Nordel study on demand participation. The 10 000 MW is an assessment of the minimum capacity on the demand side that should be able to respond to prices without discouraging loss of benefit or too high transaction costs. The Nordic drought in 2002/03 is one of the clearest examples of demand participation. Residential consumers in Norway and industrial consumers in Norway and Sweden reduced consumption significantly over a period of several months. Another important example from the Nordic market is a tight situation during a cold spell in Sweden in February 2001, during which prices spiked to unprecedented levels and peak demand was reduced with some 2% to 3%

during the critical hours compared with expected levels.[79] This made a significant difference for the total system balance. The Swedish TSO also appealed to electricity consumers to reduce load through national media, which may also have had an effect on consumption.

In the Nordic and British markets, and in PJM, significant capacity from the demand side is committed to operational reserves. Committed capacity in PJM also includes other programmes intended to promote demand participation. According to a recent review, the observed demand participation in PJM comes from similar programmes with PJM's member utilities. Responses by PJM customers included additional demand response resources adding up to a total of approximately 15% of PJM's peak demand.[80] In the British market, National Grid has been very successful at attracting the demand side in their purchase of reserves and ancillary services. The more than 4 000 MW of contracted and committed demand is for frequency and fast reserves. This market has also undertaken pilot projects with aggregation of demand for other types of reserves.

In Alberta, Canada, a liberalised and competitive market has been operational since 1998. Since then, some 3 500 MW of new generation capacity has been built in a one-price-only market. In a recent study on the demand responsiveness to electricity prices, the Alberta Department of Energy conducted a survey among the 15 to 20 largest industrial consumers. The results indicate that some 630 MW were already responsive to real-time prices, without participating in any specific programme. Among the surveyed consumers, some 800 MW was able to respond to price, representing almost 7% of peak load in 2004.[81]

Transaction costs and marginal benefit from electricity consumption set the framework for potential demand participation. This will vary with market models but also according to the characteristics of electricity consumption in different countries. How much demand participation is required to make markets robust and competitive? This depends largely on the characteristics of the electricity system and the level of competition. Researchers at George Mason University (Virginia, USA) conducted economic experiments on demand participation by letting students trade electricity in a realistic experimental environment with assumed generation and demand. In some trials, generators had market power; in others they had no market power. Some trials allowed consumers to respond to price; others did not. In the trials

79. Nordel, 2004b.

80. PJM, 2005b.

81. Alberta Department of Energy, 2005.

with demand participation, it was assumed that up to 16% of peak load could respond. Demand participation lowered prices significantly, particularly peak prices and particularly in the trials in which generators had market power.[82] In a small system in which a peak plant is the pivotal supplier, just 50 MW to 150 MW can make a big difference to the outcome. In South Australia and Victoria peak demand is projected to increase by 2.9% in 2004/05. The difference between summer peak under normal weather conditions and extreme weather conditions corresponds to 9% of peak demand. Demand participation corresponding to just 5% of peak demand would add significant flexibility to the South Australian and Victorian markets.

Planned Investment and Capacity Measures

Two of the most serious concerns in liberalised electricity markets are the threat of abuse of market power and the lack of demand participation. Effective competition and demand participation are two critical factors in establishing a robust market framework for timely and efficient investment. In several markets, these concerns have motivated specific capacity measures that provide extra incentive for investment in generation capacity. The point of departure for these measures is that the necessary level of generation capacity required to supply demand is decided centrally, for example by the system operator. This is fundamentally different from the one-price-only markets in which the necessary level of generation capacity is a matter for the decentralised decision process within the market itself.

Two approaches have been taken to implement capacity measures. One is to give a direct up-lift on the price through a capacity payment. This was used in the England & Wales pool until it was replaced with NETA in 2001. The capacity payment was computed for each hour and the point of departure for the computation was the value of lost load (VOLL), set to GBP 2000 /MWh in 2001 and thereafter set to increase with the inflation rate. For each hour, the probability of shortage was computed, based on the capacity available to generate, and the assessed peak load. The capacity payment was calculated as the VOLL multiplied by the probability of losing load. This implied that the capacity payment was low in hours with plenty of capacity surplus and high when the balance was tight. This made the capacity payment highly dynamic; it automatically adjusted to the level of investment. While theoretically ideal within a supply focused framework, it turned out to be vulnerable to gaming by dominating players. Much of the abuse of market power in the England

82. *Rassenti, Smith & Wilson, 2001.*

& Wales Pool took place in this segment of the market; withholding capacity resulted in high capacity payments. It was this gaming that triggered the change into NETA. Capacity payments are also used in Spain but the payment is fixed for all hours regardless of the actual supply and demand balance of the hour. It is adjusted on an annual basis.

The other approach to capacity measures is volume based instead of price based. Adequate generation capacity to meet projected peak demand is assessed and turned into an obligation, *i.e.* retail companies supplying consumers are obliged to contract for an amount of generation capacity that meets a certain percentage of contracted load plus a reserve. The advantage of a volume-based system is that it can be based on a cap-and-trade system, where capacity can be contracted and traded using a competitive market mechanism. PJM, the New England ISO and the New York ISO have established such capacity measures based on market mechanisms and known as installed capacity markets (ICAP). The capacity markets in these ISOs are not identical but the main principles are similar. In PJM, load serving entities are given capacity obligations and generation capacity is given credits corresponding to the installed capacity. Load serving entities must document that they have contracted a sufficient amount of credits to fulfil their obligation on a daily basis. Credits can be traded in the market, PJM organises auctions for the capacity credits and self-fulfilment is also accepted. Unfulfilled capacity obligations are penalised. In case of unfulfilled obligations, PJM purchases the missing credits at a high price, until now at a price almost 10 times the normal market price. Load serving entities that have not fulfilled the obligation are charged with this high price. The ICAP markets are accompanied with price caps of USD 1 000 /MWh in the normal energy market to limit abuse of market power.

Concerns about market power and lack of demand participation were not the only motivations for introducing ICAP obligations and markets in PJM. When utilities were unbundled and competition was introduced, incumbent generators were concerned about stranded costs. They had invested in new generation capacity to meet peak demand under the previous regulated environment. Opening up the market also implied allowing competition from neighbouring jurisdictions. The concern was that competitors, particularly those from jurisdictions that had not yet opened their markets, would be able to bid into the market at marginal costs while incumbents also had to consider recovery of invested capital in generation capacity. The solution was to guarantee recovery of invested capital through a capacity measure. The model of capacity obligations had been used in PJM before competition was introduced, so it was a natural choice.

In the northeastern United States, capacity markets face a variety of problems and challenges, and are adjusting continuously. One of the more serious problems is the implications of cross-jurisdictional trade and competition. Neighbouring jurisdictions with fundamentally different approaches to remunerating generation capacity will inevitably create some market distortions. The biggest problem has been that ICAP markets have, so far, been prone to abuse of market power. As with the England & Wales capacity payment, it seems to be too easy to game the balance of demand and supply when both are well known. Bidding volumes above the required amounts will lead to very low or even zero prices, so there is a great incentive to withhold capacity. PJM assesses the capacity market in their annual report on the state of the market. In the 2004 report, they conclude that "market power is endemic to the structure of PJM capacity markets" even though market results for 2004 were regarded to be competitive.[83] The demand side can also participate in supplying capacity credits. In PJM approximately 1% of the 3% committed demand participation as a share of peak load (mentioned in Table 5) comes from capacity credits on the demand side. The amount of demand participation in the ICAP market will, however, always be restricted to those resources that fit to the exact requirements needed to receive a capacity credit.

Functioning of capacity markets is at the centre of much research in electricity market design. Several researchers have suggested adjustments to improve competitiveness and efficiency in capacity markets. The New England ISO proposed a major reform of their capacity market that would make capacity obligations and credits more reflective of locational needs and give credits based on actual performance, rather than just availability.[84] Other researchers propose solutions that introduce more of the features used in modern financial derivates for risk hedging. A capacity obligation can instead be thought of as an obligation to contract a call option. The penalty can be thought of as the strike price of the call option, where a buyer can demand that the call option goes to delivery. Longer duration times and time-lags are also proposed to allow new investors to sell call options to help finance a plant before it is even built. The longer duration and time-lags would help new-comers curb market power.[85]

There has been theoretical research that effectively illustrates the key issues in capacity measures.[86] What research shows is that, if a price cap to curb market power is in place, a capacity market can restore efficiency compared to a one-

83. *PJM, 2005b.*
84. *Cramton & Stoft, 2005.*
85. *Oren, 2005.*
86. *Joskow & Tirole, 2004.*

price-only market by compensating for the investment incentives lost by the price cap. One of the most critical assumptions is that consumption is homogenous, meaning that all consumers and all their uses are identical in the sense of having similar profiles and similar marginal benefits from consuming. Consumers can be divided into classes but for every division a new set of regulatory instruments must be added to restore efficiency, thereby increasing complexity. Consumption is, quite clearly, not homogenous, nor can it be divided sensibly into two or three similar classes. The result is that capacity measures will induce a loss of efficiency resulting from the lack of demand participation. Capacity measures are introduced to allow price caps to curb market power, assuming that demand is in-elastic and thereby accepting a loss of efficiency. One-price-only markets aim for efficient and competitive outcomes resulting from active demand participation. What seems to divide the approaches is that, in one-price-only markets, the lack of demand participation is regarded as a transitional phenomenon that will pass with proper price signals and adjustment of market design to minimise transaction costs. In markets with capacity measures, the lack of demand participation is regarded as an inherent market failure that has to be addressed.

Contracting for operational reserves and other ancillary services could also be regarded as capacity measures in one sense and most system operators find such contracts necessary. The system operator contracts generation capacity to be ready when called upon. Plants need to be warm and ready to increase or decrease generation at short notice in the event of sudden large imbalances, even if the probability that it will be necessary to draw on them is very low. This is required to meet operational challenges but there are close linkages with the normal energy market. Payment for operational reserves becomes an element of the incentive package for new investment. Sufficient demand participation in the real-time market, which is able to respond on very short notice, should make operational reserves obsolete.

The way in which operational reserves are acquired and activated has implications for the price signals in the normal market. Operational reserves must be carefully defined and linked with the management of real contingencies. It should be as easy as possible for the demand side to participate. The reserves should be bought using a competitive process to ensure that when sufficient real-time demand participation makes them obsolete, competition will drive the contract price to zero. Operational reserves should only be activated in case of real contingencies. Loosely defined operational reserves with loosely defined rules for activation can severely distort investment incentives. If the need for operational reserves is assessed from a normal evaluation of the balance between demand and supply, contracting of operational reserves will slide into a normal capacity measure. If operational reserves are activated at low prices during normal tight situations, where there are no contingencies, they will dampen the market price that should have signalled the tight situation. As a consequence, investments will not be made and the TSO is forced to further increase acquired operational reserves. They are no longer operational reserves but, simply, reserves.

Active demand participation and market power remain serious concerns in all markets which could be a good reason to introduce capacity measures. On the other hand, capacity measures have proven to be prone to market power themselves and one-price-only markets have so far not failed to deliver. Competitive one-price-only markets have not led to excessive market power and evidence so far indicate that clear and transparent cost-reflective price signals do give incentives for timely and adequate investment. There is some evidence emerging indicating that demand will, indeed, participate if the framework is right and when consumers have something to respond to, *i.e.* uncapped prices. The merits of capacity measures still needs to be proven and the failure of one-price-only markets have still not materialised. This suggests that capacity measures should be avoided, particularly if the general policy approach is only to regulate the industry when necessary.

Capacity measures are still at the heart of the debate in North America, Europe and Asia. This raises additional issues concerning the co-existence of various approaches to incentivising investment. In a EU directive concerning security of electricity supply soon expected to enter into force, it is proposed that EU member states can use both a one-price-only-market and an obligation-based measure (a capacity measure) to maintain balance between electricity supply and demand.[87] Regardless of the individual merits of the two approaches, their co-existence raises issues of efficient incentives for investment in an internal market and free-riding across borders. Capacity incentivised through special capacity measures in one jurisdiction will distort investment incentives in neighbouring jurisdictions with one-price-only markets. In relation to the development of an internal market on the Iberian Peninsula, the market implications of the Spanish capacity payment is one of the critical issues. This could call for a stronger role for the European Union in guaranteeing consistency across the intended internal European market.

87. *European Parliament, 2005.*

Investment in Transmission Networks

Effective performance of transmission networks shapes the operation and development of efficient wholesale and retail electricity markets, particularly the emergence of effective regional markets. Effective performance of transmission networks also underpins efficient development of inter-regional trade, which contributes to effective competition and reduces scope for market power abuse. It also improves capacity utilisation (both generation and networks) and helps to defer expensive investment in generating capacity. In addition, more effective regulation reduces excess transmission capacity. As a result, transmission systems are becoming more stressed and prone to congestion. Other policy priorities have led to increased use of less greenhouse-intense forms of distributed and intermittent generation, making the operation of transmission networks increasingly complex and expensive. At the same time, more stringent environmental approval processes are making timely network investments more difficult to realise.

The combination of these factors creates new challenges for maintaining reliable transmission services and maximising transmission network performance. Transmission owners and operators have far less certainty about the demands that will be placed on their networks and less capacity to undertake integrated planning and development of transmission networks as a whole. Transmission system operators' capacity to manage system balancing and efficient network development is, thus, greatly reduced, particularly within regional markets incorporating multiple owners and operators. At the same time, persistent concerns about regulatory and policy uncertainty hold the potential to magnify the business risks faced by transmission network owners, with the potential to discourage investment and encourage more conservative network performance. The situation can be further complicated when vested interests serve to weaken incentives for efficient network performance and where jurisdictional boundaries unduly affect the emergence of efficient market structures or design.

The rate of investment in transmission systems appears to be slowing, particularly in North America and Europe, and especially in jurisdictions with liberalised electricity markets and reformed economic regulation. IEA projections of the need for investment in transmission systems in OECD countries are USD 498 billion for the period 2003-30, corresponding to roughly 25% of total investment needs in the electricity sector. There is much focus on the role of interconnections to enhance competition and the efficient use of resources. In 2002, the European Union even adopted a target for minimum transmission capacity on interconnectors between member states, to support the development of integrated and reliable regional electricity markets. The priority projects pointed out by the European Commission in its TEN-E project can be seen in connection with those targets.[88] The European Commission projects that the total costs of the interconnections across the prioritised bottlenecks will be some EUR 5 billion until 2013. IEA projections for transmission investments in the European Union between 2001 and 2010 are EUR 49 billion, illustrating that investment in transmission interconnectors to relieve important congestions is only a share of total required investment.

Following a surge in investments in new transmission lines in the 1960s and 1970s, the increase has slowed down. Figure 16 shows the annual increase in lengths of transmission lines in selected European countries.

Figure 16

Annual average increase in length of 220-400 kV transmission lines in 16 European countries

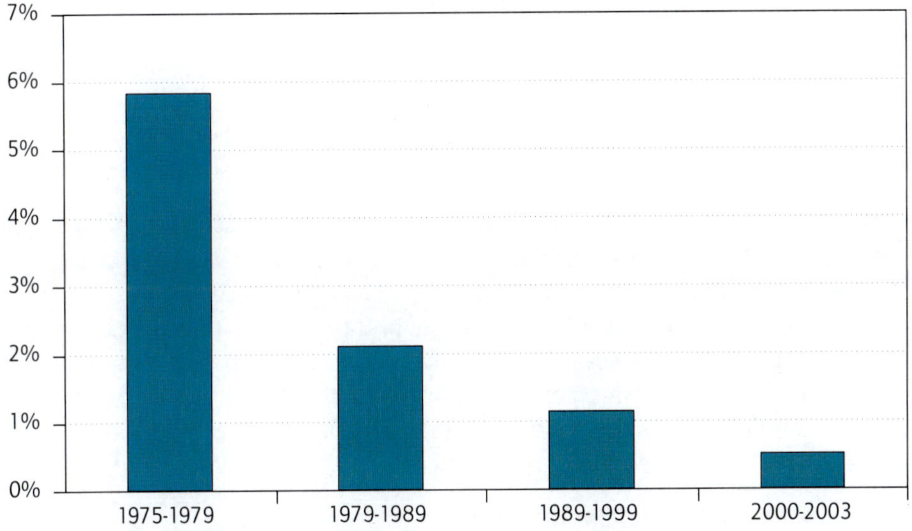

Source: UCTE and Nordel.

A similar development has occurred in the United States.[89] Transmission investments in the United States fell to unprecedented low levels in 1997 and 1998 but have now recovered to levels seen in the early 1980s in terms of USD real.

Investment in transmission networks has also become an important policy issue. It has been suggested that chronic underinvestment in transmission capacity, especially interconnector capacity, was an important contributing

88. *European Commission, 2004a.*
89. *IEA, 2002.*

factor leading to the cascading blackouts experienced in 2003 in Europe and North America. Thus, reinforcement of the main transmission network ought to be a top priority to ensure reliability and efficient market development. Investment in new transmission lines could add strength and resilience to transmission systems, enabling them to face contingency events. But such investments may have the character of gold-plating. Changes in the way transmission assets are maintained and systems and markets are operated may be a far more cost-effective way to enhance system security. This issue is discussed thoroughly in a recent IEA publication.[90]

Concerns about under-investment in transmission network infrastructure raise the issue of whether electricity market reform, particularly structural reform of the electricity sector and reformed regulatory arrangements, create undue barriers to efficiently timed and sized development of transmission networks.

Recent investment trends may be partially explained by pre-existing excess capacity in transmission networks and more effective economic regulation, which introduced new financial and regulatory disciplines on transmission investment projects and decision-making processes. The question now is whether regulators have struck the right balance between passing efficiency dividends to consumers and ensuring adequate returns for efficient transmission projects.

The challenge of investment in transmission in liberalised markets is to exploit the price signals for efficiency that effective markets produce. At the same time, effective regulation must be maintained that properly accounts for cost-effectiveness, reliability concerns and the role of transmission interconnectors in enhancing competitiveness. This all takes place during a period during which electricity systems are changing fundamentally; liberalised electricity markets are in a transition and, due to insufficient unbundling, vested interests are still affecting the decision process.

The goal of market design that creates cost-reflective and transparent locational price signals is to set a price on the value of transmission capacity. With a fine tuned approach, such as nodal pricing, the signals are precise; with a zonal approach, they are less precise. Appropriate market design is a key parameter in creating incentives for efficient transmission investment, particularly for interconnectors. For example, inappropriate regional pricing models that unduly mask intra-regional transmission congestion, or do not provide access to accurate and timely information about the performance of transmission networks, can increase investment risks and create a potential barrier for alternative investments to alleviate congestion.

---

90. IEA, 2005.

Some market models and regulatory arrangements may also create perverse investment incentives, reflecting inherent conflicts of interest. For example, it is possible that arrangements allowing network owners to retain congestion rents resulting from congestion on interconnectors may encourage network owners and operators to refrain from efficient investments to alleviate transmission bottlenecks in an attempt to maximise rents. Such arrangements may encourage network owners to transfer the effect of intra-regional congestion to network boundaries, also in an effort to capture additional congestion rents and defer costs of managing internal congestion through re-dispatch. In some cases, ineffective unbundling can encourage transmission network owners to discriminate in favour of incumbent generators - at the expense of efficient network investments to maximise trade and competition. Such behaviour would not only discourage efficient network investment but may have the potential to distort efficient operation and development of regional electricity markets.

One approach to investment in transmission interconnections is to simply let it compete with generation on equal terms. Interconnections are, after all, the alternative to generation investment. The concept of merchant interconnectors competing with generation, demand response and other network owners to deliver services across regional electricity markets would be financed by congestion rents. An investor in a merchant interconnector would size and locate the investment so that price differences between the two ends of the interconnector multiplied by flow can finance the project. Greater reliance on competitive transmission services, based on competition between merchant lines, may support more effective use of price signals to strengthen incentives for efficient transmission network performance. Such an approach may be more responsive to electricity markets, particularly in the context of inter-regional congestion management. It may also help to strengthen market-driven signals for efficiently timed, sized and located new investment, supporting inter-regional trade. Together, these factors may help increase the economic benefits derived from electricity market liberalisation and allow a more coherent and transparent approach to managing the commercial risks associated with transmission network operation and development in competitive electricity markets.

There are a number of real-world challenges to this concept, however. Transmission investment involves economies of scale and high fixed costs. This may imply that the most efficient investment response to inter-regional congestion could completely eliminate physical constraints and the related price differentials between regions or nodes, thus removing the underlying cash flows required to fund investment. Development of transmission systems is also driven by reliability requirements and such requirements can only be ensured as a regulated investment. Transmission investment motivated by reliability and system security requirements will still compete with merchant lines and generation. This adds considerable regulatory uncertainty to merchant investments.

Several merchant interconnectors currently operate in Europe, North America and Australia. Those commissioned in the era preceding electricity market liberalisation tend to operate in a manner that raises concerns about the potential for merchant interconnectors to support market power abuse. However, evidence from Australian NEM, suggests that merchant interconnectors responding to market price signals could provide an efficient means of helping to integrate network services with competitive electricity markets. The Australian example also highlights some of the challenges. Two merchant projects were completed in Australia at the outset of the market when the need for interconnections was relatively clear and seemed quite certain to benefit from interlinking diverse generation portfolios. Directlink was completed in 2000, connecting Queensland and New South Wales. The Murray link was completed in 2002, connecting Victoria and South Australia. In addition, a regulated transmission project was proposed to connect New South Wales and South Australia by TransGrid, the transmission company owned by the NSW government. The profitability of this project changed fundamentally with Murray link, which illustrates some of the challenges in creating a regulatory framework that enables the co-existence of merchant lines and regulated lines in the same system.[91] Regulation easily fails to create a sufficiently stable and predictable business environment to create efficient incentives for merchant lines. Murray link has since been transformed into a regulated interconnection.

To date, transmission networks have typically been treated as natural monopolies that should be physically separated from the other components of the value chain and separately regulated. Prior to the introduction of competition reform, transmission network regulation was typically based on some form of cost-plus methodology, which allowed network owners to pass through to customers all costs approved by the regulator. This form of regulation provided little incentive for efficient operation or investment and, in some cases, allowed network owners to pass on the risk and cost of poor investment decisions and inefficient operations.

Electricity market reform is generally accompanied by regulatory reform. Non discriminatory access provisions were introduced to facilitate efficient

91. Littlechild, 2004.

competition and trade. At the same time, incentive regulation, typically built around CPI-X price or revenue caps, was introduced to provide a quasi-competitive discipline to encourage more efficient operation and development of transmission networks.

There is some concern that regulators are too focused on reducing operating costs, at the expense of incentives for effective operational performance and efficient maintenance and investment. This can lead to operational decisions with substantial efficiency losses and financial costs for consumers and market participants. Potential examples include removing major transmission lines from service for maintenance during peak periods, inadequate maintenance and investment in transmission networks, and transferring intra-regional congestions to interconnectors.

Creating regulation for efficient transmission networks is inherently challenging, particularly in terms of properly integrating risk management and asset availability in the regulatory incentive package. Transmission maintenance planning should be conducted with the costs and benefits expressed in the electricity market in mind. Risk management is also, to a large extent, connected to the discussion of firmness of financial transmission rights (FTRs), as discussed in Chapter 3. If network users are able to buy firm FTRs, their value increases as a risk management tool for market participants. Utilisation and, hence, value may be further enhanced if transmission service users are able to make a contract one day, one month or even one year ahead. On the other hand, the length of the commitment also increases the risk of not being able to fulfil the contract. This comes at a cost for the body, often the system operator, who underwrites the contract due to the consequent need to re-dispatch generation. FTRs can only be firm if those bodies taking the risk, such as transmission owners or system operators, are able to manage the risk. In the Nordic market, all users share the risk. The Nordic TSOs are allowed to pass on all costs of re-dispatch to grid users for firmness committed one day-ahead and can use congestion rents to pay the cost of re-dispatch. Collection of congestion rents and passing on costs of re-dispatch are two approaches that can be used in the incentive package to ensure efficient utilisation.

There is much debate about firm FTRs. Some traders advocate for annual and monthly firm FTRs; some network owners point out that they are not willing to bear the risks of such commitment. An owner of a commercial and competitive merchant interconnector would find the commercially sensible balance. In an environment of regulated transmission ownership and vested interests, risk management of transmission assets must be an integrated part of the regulatory framework, particularly if the independence of TSOs is questionable. In the England & Wales market, a regulatory change in 1994 made re-dispatch costs a part of the incentive package for National Grid. The availability of transmission assets improved markedly and re-dispatch costs decreased correspondingly.

Uncertainty may be magnified in the case of transmission network investments designed to reinforce network reliability. Emerging regional markets have the potential to fracture responsibility for maintaining reliability across many network owners and operators. Although regulations may require individual transmission owners/operators to meet certain minimum reliability and safety standards, they are not usually directly exposed to the financial consequences of catastrophic system failure in electricity markets.

Regulated processes to assess proposals for reliability augmentations have proven problematic in practice. Reliability-based investment proposals, like other regulated network augmentations, are typically developed by network owners and subject to regulatory approval. However, the regulator can be at a considerable disadvantage in assessing such proposals, particularly where it is reliant on information submitted by the network service provider that is proposing the investment. Cost-benefit assessment in these circumstances can be problematic and open to dispute, possibly leading to delays and uncertainty. Access to accurate and reliable information about the operational condition of the network is crucial to help improve the effectiveness, timing and credibility of such processes. Some regulators are exploring options to improve assessment processes, including the use of probabilistic analysis to analyse the potential financial implications of network failure on electricity markets. Probabilistic assessment is currently used in PJM and in Australia (VECorp), and has been proposed in New Zealand.

New transmission network investment is also affected by other approval processes including construction, siting and environmental approvals. Greater environmental sensitivity, combined with increasing local community opposition, sometimes slows approval processes, making licenses more difficult and time-consuming to obtain. Inefficient and inconsistent approval processes can create uncertainty and regulatory risk, which may undermine timely and appropriately sized investment responses.

To date, incentive regulation has been very successful at driving down transmission network operating costs and passing savings through to consumers, usually in the form of lower tariffs. However, concerns are emerging about the wider operational and investment incentives created by regulation. Underinvestment in transmission networks and interconnectors has been raised as a critical issue in the United States and appears to be an emerging issue in Europe.

A particular challenge for regulators is to achieve an appropriate balance between incentives to minimise costs and network charges on the one hand, and augmentation of transmission networks on the other. This challenge is likely to be magnified by the inherent information asymmetry associated with regulation of transmission network services. In Australia, a newly implemented regulatory test assesses costs and benefits, both in terms of enhanced trade and system security.

Experience with economic regulation is also raising concerns about regulatory uncertainty, which principally relates to the amount of discretion afforded regulators in their interpretation and application of regulatory provisions. One suggested solution is to tighten legislative frameworks to reign in regulatory discretion by clearly prescribing the nature, scope and limits of discretionary powers, thereby minimising uncertainty and regulatory risk that may otherwise have the potential to hinder efficient market operation and development. This may prove a substantial challenge in practice, especially where the boundaries of regulatory and policy responsibility may be overlapping and somewhat ambiguous. Experience to date suggests that, in practice, economic regulation in electricity cannot be simply divided between policy and administration. Issues of detail matter in electricity markets, and regulatory interpretation and decisions on matters of detail can critically influence strategic policy outcomes. Transparent, consistent and balanced regulatory decision-making and outcomes are required to help address perceptions of regulatory risk and uncertainty.

Challenges associated with minimising regulatory risk and uncertainty are greatly magnified in regional markets that span several jurisdictions. Typically, policymakers have sought to minimise these risks by adopting consistent regulatory rules across jurisdictions within regional markets. However, concerns about the potential for inconsistent interpretation and application of regulatory provisions amongst different regulators within regional markets remain and continue to create regulatory risk and uncertainty, particularly in relation to the operation and development of interconnectors linking different jurisdictions within regional markets.

Policymakers and regulators are taking steps to address these risks. For example, regulators in Europe, North America and Australia have created multi-lateral fora to pursue greater co-ordination and harmonisation of regulatory interpretation and administration. Some regulators sought to improve transparency and certainty in relation to regulatory interpretation and application by issuing non-binding statements of regulatory intent. Some jurisdictions have gone further by consolidating regulatory functions across several jurisdictions into a single regional regulatory body that reports to member governments collectively. A recent example is the new National Energy Regulator established by the governments participating in the Australian National Electricity Market.

Co-ordination of Transmission Investment

In a regime of vertically integrated utilities, transmission investment is an integrated part of an overarching investment plan. The need for transmission lines can be carefully incorporated into a plan for investment in generation capacity. System security issues are a priority concern, particularly when considering development of interconnectors between jurisdictions. When responsibilities are unbundled in liberalised markets it raises questions about the efficacy and compatibility of such planning regimes. This is particularly true in a competitive market environment characterised by independent, decentralised decision-making: network owners and operators can no longer accurately predict or control key decisions determining future network use, especially in larger regional markets. It also raises practical issues associated with coordinating planning processes in larger regional markets where responsibility for planning is shared across several network owners/operators. Furthermore, existing processes may unduly favour investment in network solutions over potentially more efficient generation and demand response options, reflecting the vested interest of transmission owners/operators responsible for planning processes. Regulators responsible for managing this risk may be at a particular disadvantage in attempting to moderate this behaviour in practice, particularly when they must rely on information and technical advice provided by incumbent network owners/operators who have a pecuniary interest in the outcome. This could prove to be a particularly difficult issue in the context of assessing reliability-based investment proposals.

Fractured ownership responsibility for transmission networks across regional markets may also affect incentives for transmission network investment. Where there are many transmission network owners and operators within a single regional market, it is unlikely that any of them will have a clear incentive to undertake transmission network development from a market-wide perspective. Unclear or shared ownership responsibilities, combined with practical challenges of coordinating investments amongst different network owners, have the potential to weaken incentives and increase risks, particularly for interconnector investments. Regulatory uncertainty may increase these risks, especially if construction approvals and regulated returns are subject to different regulatory processes and decisions on either side of an interconnector – which are likely to be further magnified where interconnectors cross national boundaries.

Greater coordination becomes essential in liberalised markets, especially amongst network owners and operators within larger regional markets, to ensure that planning processes reflect a more holistic and regionally integrated perspective.

After experiencing some of the challenges these issues raise, several markets took the approach of transforming planning processes into a tool for accurately and transparently informing market participants and regulators about transmission network use and resources, as well as emerging trends. The intent is to improve independent, decentralised decision making within electricity markets and to support more effective regulatory assessment of transmission network investment proposals. The annual Statement of Opportunity published by NEMMCO for the Australian National Electricity Market is one such example. As a response to a NEM review in 2002, NEMMCO extended its Statement of Opportunity with an Annual National Transmission Statement (ANTS). The Statement of Opportunity carefully describes and analyses demand and supply fundamentals. The ANTS analyses national transmission flow paths, forecasts interconnector constraints and identifies options to relieve constraints. It is then up to market participants, including network owners, to respond to the needs by adding either generation or transmission assets.

In the Nordic market, the opportunities, challenges and caveats of transmission planning in liberalised markets is emphasised in the co-ordinated transmission plans developed by Nordel. All Nordic TSOs have a long history of transmission planning. Now Nordel has developed a co-ordinated plan in which important congestion points in the Nordic transmission system are identified and assessed by modelling the Nordic market. Economic modelling has been used to assess the value of relieving single congestion points. A characteristic of the Nordic market is that large hydro-electric resources, located primarily in the north, are connected with demand and thermal generation plants in the south and on the European continent. In such a system, relieving one congestion point often just has the effect that the congestion is moved to another point without benefiting substantially from the added transmission capacity. The economic modelling used in the Nordel transmission planning identified flow paths or series of bottlenecks. Benefits of relieving congestion points and paths are assessed for both generators and consumers, and for both the whole Nordic region and country by country. This points to some of the political challenges of transmission investment and again stresses the importance of impartiality and independence from incumbent generators to enable efficient transmission investment. The Nordel transmission planning exercise of 2004 identified five major transmission investment projects. Some will create losses for generators in some Nordic countries and for consumers in other Nordic countries, but in aggregate they will add considerable economic benefits to the joint Nordic economy, also outweighing the assessed investment costs of approximately EUR 1 billion. A process towards final project decision, including financing the investments, is currently evolving.

WHEN DO LIBERALISED ELECTRICITY MARKETS FAIL?

The perfect market is a theoretical ideal rather than an achievable goal. No markets are perfect and all markets are in constant evolution. Competition is rarely perfect. Policy and regulation change. Technology and taste change. Innovation pushes development forward in steps that can create sudden big changes. Electricity markets are no different in that sense. No electricity market is perfect, or is ever likely to be so. It is more relevant to assess the performance of liberalised electricity markets by comparing against the alternatives. Does a less-than-perfect liberalised electricity market perform better than a less-than-perfect vertically integrated and regulated system?

There are some serious concerns in liberalised electricity markets. Competition and cost-reflective pricing does not come easy; it can only be accomplished with a high level of commitment by policy and decision makers. Even with the necessary commitment and determination, markets will still take a long time to prepare and even longer to be fully implemented and developed. Electricity market liberalisation is a process. With that in mind, are there still some factors in liberalised markets that simply cannot be managed in a satisfying way? Are there some factors in the economic optimisation of markets that will never be taken into account, regardless of how competitive these markets are and how well they are designed? It has been argued that such so-called "market failures" are inherent in electricity markets and can be found in many parts of the value chain.

For example, some argue that electricity markets are inherently uncompetitive: economies of scale will inevitably lead to large companies that will have the power to collude. This claim seems to contradict the recent development of more varied and more flexible units of generation technologies, including combined cycle gas turbines (CCGTs). If large economies of scale are apparent in a market, it may rather suggest that regulation of the market failed to remove the barriers that allow small companies to enter and operate. Another claim is that the demand side is, in fact, inherently inelastic – that demand-side participation cannot be expected, and that is a serious market failure. Considering that all uses of electricity will have different values for all users, the thought that all demand is inelastic is certainly counter-intuitive. In the end, it may not be possible to lower transaction costs sufficiently to make sufficient demand participation profitable, but given the high focus on supply in the current market organisation this conclusion seems premature. In any case, innovation and technology are constantly lowering barriers. The danger is that efforts to involve the demand side may fail if markets are designed assuming that demand is inherently inelastic.

Another concern is the prospects of research and development in liberalised electricity markets, when the priorities of market players change with the emergence of competitive pressures. To date, private commercial companies have not undertaken sufficient R&D in areas that are mainly of public interest, such as renewable energy and energy efficiency. Experience shows that R&D requires policy attention within areas of public priority, as is required in non-energy sectors when markets fail. That being said, one general lesson from liberalisation processes in other sectors is that competition also creates incentives for active innovation.

In two areas, alleged market failures in the electricity sector seem more related to the inherent characteristics of electricity rather than the result of underlying regulatory failures. Electricity is a unique product. Modern society is increasingly dependant on electricity so the demand for reliability of supply is high. At the same time, it is a highly complex product mainly due to the fact that it cannot be stored. How unique is electricity and at what point might markets fail to secure a reliable supply? Also, electricity generation with most conventional technologies has impacts on the environment and the climate, yet markets do not properly account for external environmental costs - a classical market failure. With policies now in place to address external environmental effects, what are the implications for liberalised electricity markets and how might liberalised electricity markets contribute to meeting new challenges such as climate change?

When they arise, real market failures due to the inherent characteristics of electricity must be addressed with policy measures that try to compensate for the failure or internalise the costs. Again, the appropriate policy instrument may not be found by referring to some theoretic ideal, but rather by comparing alternatives. Some market failures may leave the electricity market less than perfect and, thus, policy options designed to address the failures should be assessed on their ability to improve the overall outcome.

Reliability of Supply in Liberalised Electricity Markets ......

In liberalised electricity markets, the intent of price signals is to direct all actions toward efficient outcomes. Thus, it is reasonable to ask: Are there any points along the value chain - from electricity generation to transport to consumption - at which price signals can be expected to be insufficient? Are there any parts of the value chain in which markets fail to produce efficient outcomes? The value chain can be broken up into three parts, each of which addresses different critical components of a reliable supply of electricity (Figure 16). Another way to put it is that the concept of reliability of supply needs to be "unbundled" to be analysed in an unbundled sector.

Figure 17

The value chain for reliable electricity supply: Energy security, adequacy and system security

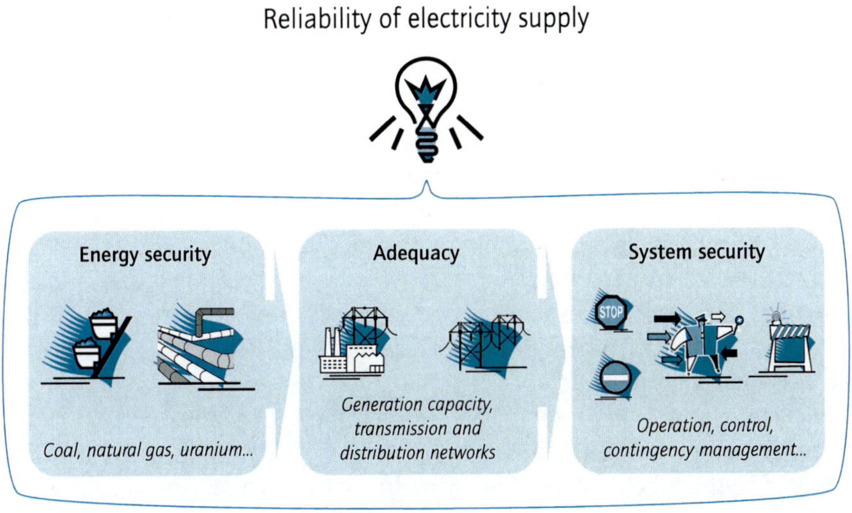

First of all, there must be a secure fuel supply as input for power generation. Secondly, adequate generation and network assets are needed to transform the fuel into electricity and transport it to the consumer. Finally, because electricity cannot be stored, it is a significant challenge to operate the system securely – to actually keep the lights on. In which parts of this value chain are markets likely to fail?

Energy Security of Supply for Electricity Generation

Security of upstream energy supply for power generation is a matter of how well fuel markets function. Electricity is mainly produced from coal, natural gas, uranium and hydropower. Other renewable energies are attracting more and more attention. While oil was a major fuel for electricity generation until the oil crises of the 1970s, its share has since decreased to insignificant levels in most countries. The global coal market is competitive, resources are abundant and it can easily be stored. Uranium can also easily be stored. Considering the minor role of oil in electricity generation the focus ends with natural gas. Natural gas resources are relatively concentrated in few countries, storage is relatively costly and it has been grid bound even if this constraint is changing with the development of liquefied natural gas (LNG). Natural gas has an increasing importance in electricity generation. Due to their low capital costs and short construction times, combined cycle gas turbines (CCGTs) are now a preferred technology in many situations in many liberalising electricity markets. Natural gas also has an environmental comparative advantage to coal with its lower emission of greenhouse gases per generated MWh of electricity in CCGT use.

Markets for natural gas are also liberalised or liberalising in several IEA member countries. In the European Union, natural gas markets are liberalising in a process that parallels electricity market liberalisation, even though gas market liberalisation lags behind somewhat. Natural gas markets are also under strong influence from the development of liquefied natural gas (LNG), which can potentially add significant flexibility to regional and global gas markets. A recent IEA study on security of gas supply in open markets describes the status of gas markets and discusses the role of LNG.[92] As is the case with electricity, liberalisation of gas markets is a long and complex process; in several IEA member countries there are still many challenges to overcome. Natural gas reserves are concentrated in relatively few non-IEA member countries; this raises considerations of geo-political dimensions concerning the functioning of the upstream part of natural gas markets. Even though this concentration raises concerns similar to those associated with oil reserves, one major difference is that natural gas is only one of several fuel options for power generation. The possibilities for substitution of fuel for electricity generation in the mid-term are much greater than is the case for substituting oil with other alternatives for transportation.

Clearly, there are reasons for concern regarding the security of gas supply: will sufficient investments be made upstream, will markets be successfully liberalised down stream and will LNG reduce dependence of specific suppliers and increase competition? In relation to liberalised electricity markets, the issue is whether investors fail to take security of gas supply considerations into account. CCGTs are a preferred technology, but the profitability of a CCGT investment will depend on a secure supply of natural gas which is competitively priced. At present, lack of market liberalisation for natural gas is

92. *IEA, 2004a.*

regarded as a barrier for investment in CCGTs.[93] Investors in CCGTs consider the full natural gas value chain, from well-head to generation plant. Some electric utilities choose to enter the natural gas business as a part of an overall strategy that also includes security of gas supply considerations.[94]

If investors fail to fully account for the economic consequences of gas supply disruptions or market manipulation - considering both probabilities and costs - policy intervention may be appropriate. A potential risk is that the total electricity generation system could become too dependent on natural gas through increased use of CCGTs. Two main policy options have been applied to address concerns for gas supply security: one is to promote alternative technologies to ensure diversification; the other is to focus on the natural gas market itself. Policy makers can improve the functioning of the down-stream gas market through liberalisation and regulation. Dialogue with supplying countries can help address up-stream issues. Application procedures for LNG and other gas infrastructure investment projects can be improved.

A policy to promote alternative technologies spills directly over into liberalised electricity markets. Direct subsidies to promote specific alternative technologies or regulatory barriers to build CCGTs are intended to distort investment signals. Distortions implemented in one country will affect internal electricity markets in neighbouring countries. Are such distortions warranted and efficient? The CPB, the Netherlands Bureau for Economic Policy Analysis, is an independent governmental agency that conducts economic policy analysis. It recently conducted a cost-benefit analysis on various aspects of reliability of supply, including the case for policy intervention to decrease the Netherlands' dependence on natural gas for power generation. Considering assessed probabilities, costs and benefits in relation to gas supply interruptions or market manipulation intervention could not be justified.[95] Recognising that secure supply of natural gas is a great concern for investors in CCGTs and that intervention based on judgements by governments are not likely improving decisions alternatively made by commercial investors it seems more relevant for governments to focus on minimising regulatory uncertainty in the gas market.

Adequacy of Generation Capacity and Transmission and Distribution Networks

Whether there is adequate generation capacity to meet all demand at all times depends on whether investments are made at sufficient volumes, at the

93. *Speckler, 2005.*
94. *de Tomás, 2005.*
95. *CPB, 2004.*

right location and in a timely fashion. There is no reason to believe that competitive electricity markets cannot provide incentives for timely and efficient investments. The key requirements are: real competition (including regulated third-party access to unbundled networks); cost-reflective locational pricing; and market rules that ensure transparency and low transaction costs. In addition, there is the regulatory challenge of creating a smooth regulatory framework for transparent and clear approval processes of new generation plants. At present, market players are investing in liberalised electricity markets, even without additional capacity measures.

Networks tend to be natural monopolies. Markets will, in general, fail to provide signals for adequate and timely investment. Investment in networks depends largely on the regulatory framework and the incentives incorporated therein. Thus, liberalisation does not, in principle, change the situation for networks - except perhaps if they are privatised as part of the liberalisation process. However, liberalisation does increase the focus on network operating costs, and many countries have introduced models for regulation of network tariffs with strong incentives for cost cutting. This high initial focus on costs raised concerns for the quality aspect in some countries, particularly in the sense that quality is directly linked to adequacy. In several countries, models for regulating network tariffs are now being adjusted to also include direct incentives for quality - and, thereby, incentives for investment.

Markets do not necessarily fail to provide signals for adequate investment in transmission networks. In fact, locational pricing should create incentives for investment in a way that treats generation and transmission assets as substitutes for one another. So far, this is one of the disappointing failures of liberalised electricity markets. Almost all investment in transmission networks remains regulated; there are only few on-going examples of so-called "merchant lines".

System Security in Transmission System Operation

The value of electricity to its consumers is determined by time, location, volume and quality. Liberalised electricity markets can set a price in terms of the first three factors, but quality is so far a uniform product delivered to all consumers, regardless of price signals. Technically and economically, it is still not possible to deliver electricity at varying quality, depending on the willingness to pay. All consumers receive electricity at a certain standard quality; if problems arise that force the system operator to shed load, this will happen more or less randomly without taking notice of willingness to pay. In this context, load shedding is thought of as load that is cut involuntarily after all other options have been exploited including voluntary load reduction contracts with consumers. A full blackout happens to everybody. Quality is also a characteristic of electricity supply that everybody can benefit from without reducing the benefit for others. Just because someone pays, does not mean he will get a higher quality product or have a lower probability of being disconnected. The case is the same for consumers who decide not to pay for quality, *i.e.* there is an incentive to try to free ride. Consumers cannot claim individual property rights to quality of electricity supply, which makes it a classical public good. In general, markets fail to balance supply and demand for public goods at a level that makes everybody happy. Without intervention from policy makers, markets will deliver public goods, such as electricity quality, at a level that is lower than consumers would prefer.

To date, establishing independent system operators is the primary means of addressing the failure of liberalised and competitive electricity markets to deliver quality. These system operators ensure quality by balancing supply and demand in real-time operation. They manage the interface between electricity as a normal private good traded in a market and the public good aspect of the product related to the quality.

System operators have a large toolbox from which to draw instruments: balancing real-time supply and demand using real-time markets and other ancillary services; a fixed set of reliability criteria that determines the limits of the transmission system; immediate and transparent disclosure of critical information to the market place; skilled personnel and equipment to carefully monitor and control the system, etc. Many of these instruments have close links to electricity as a private good traded in a market. If they are not used according to very strict standards - and in very disciplined way - they will blur the interface between electricity as a private and public good. Opaque and non-transparent definition and management of congestion blur price signals: markets will respond less comprehensively and precisely, and the public good sphere of electricity will increase.

In 2003/04, there were some remarkable blackouts and incidents related to system security. The most publicised events included blackouts in North America, Italy, southern Sweden, and eastern Denmark. An important incident also occurred in Australia, but was managed well enough to avoid a major blackout. In each of these cases, official investigations do not blame market liberalisation. They did, however, raise some issues concerning system security in competitive markets that must be better addressed in the future.

Liberalised markets significantly increased cross-border trade, in terms of both volume and variability. These exchanges contribute to the overall benefits of liberalised markets. However, they must be carefully managed by system operators to avoid becoming a threat to system security. The above incidents stressed the importance of co-ordination and co-operation between system operators in general. Initially, this was directed primarily through voluntary agreements in North America and Europe. After the blackouts, the United States implemented legislative changes and UCTE made their operational handbook binding. The incidents pointed to the importance of monitoring the actual fulfilment of cross-border agreements. The question is whether compliance monitoring can be effective unless these agreements are made legally binding.

In reality, reliability criteria extend far beyond the role of system operators. In the North American and Italian blackouts, insufficient tree trimming along power lines played an important role in the actual failure of transmission systems. This raises the issue of including a quality assurance dimension to regulation of transmission systems, as a means of ensuring good and secure availability of transmission lines. As is evident, this relates to sophisticated regulation rather than the impact of liberalisation. A recent IEA book that focuses on these case studies also reviews and discusses the issues of transmission system security in competitive electricity markets.[96] Its main conclusion is that liberalisation fundamentally changed the usage of transmission systems and management – and that operation of transmission systems still needs to adapt to these changes.

What is most important is that the much-publicised blackouts and the day-to-day management of system operation in liberalised markets do not point to unsolvable issues. They do make it clear that management in all parts of the value chain – from generation to consumption – must be re-considered in liberalised markets, including all aspects of system operation. Very few of the issues raised with liberalisation relate to specific problems from the unbundled structures in liberalised markets. Almost all of the issues relate to aspects that would improve system operation in terms of security and efficiency, regardless of the organisation of the market. For example, cross-border trade improves the efficiency of the overall system and the usage of transmission assets. It changes the usage of the system itself in a way that forces it closer to its limits, which makes the need for reform of system operation principles more pressing. Well-managed co-operation between system operators is not a threat to system security but rather an important potential for improved security. Well co-ordinated system operation across several systems should make all systems stronger than each would have been on its own. Policy intervention from

96. *IEA, 2005c.*

governments may be needed to secure these obligations and to ensure that market players and independent system operators are taking steps to achieve these aims. If players are not co-operating and co-ordinating their efforts, governments must take policy action.

Addressing Environmental Issues and Climate Change .....

The environment and climate are classical public goods - to the extent that markets fail to account for the costs and benefits associated with activities that have an impact upon them. There is no reason to expect that liberalised markets will voluntarily begin to account for external environmental effects. This will only happen through policy intervention. In theory, it is straightforward to internalise an external environmental cost. The damaging activity can be penalised through a tax that corresponds to the cost such activity inflicts on society. However, environmental costs are very rarely computable in a way that is precise and can be commonly agreed upon. This raises a whole range of issues that will interfere with markets, including questions of international competitiveness in open markets.

Environmental costs that are borne only by generators and only in some countries will give competitive advantages to economies in countries that do not consider these costs. This relates not only to competition in electricity but also transfers to the competitiveness of products produced using electricity. In that sense, differing approaches to the internalisation of environmental costs distort both liberalised electricity markets *and* all other global markets, particularly those with high electricity content. The issue of how to assess and internalise environmental costs is an issue for liberalised electricity markets, as it is an issue for other open markets. At present, the issue remains beleaguered by international disagreement. Again, it is more relevant to consider policy options with reference to the alternatives, rather than using some theoretic perfection as the benchmark.

Generating electricity creates consequences for the environment and the climate. These should be addressed by all governments through policy initiatives. The key question for liberalised electricity markets is how various policy options will affect the market functioning and its outcomes. Policy initiatives are intended to create the exact effect warranted by the external cost. But it is clear that various policy options have different distortive effects in a liberalised electricity market. This is particularly true in regional markets in which environmental costs are addressed differently in different countries. The fundamental question is this: are there policy options that undermine the way liberalised markets intend to create efficient outcomes, in terms of both operation and investment?

Among the policy options considered and implemented in liberalised markets two general types stand out as having the highest impact: those that support, promote and subsidise specific technologies and those that impose a cost on damaging activities. The first includes R&D programmes for renewable energy and other initiatives to ease the construction of renewable energy generation plants. Such initiatives are common in many other sectors and can hardly be seen as detrimental for the functioning of liberalised electricity markets, as long as they do not involve large-scale deployment. Different countries have different requirements regarding accepted standards for generation plants. These affect the costs of generation plants and distort international competition. On the other hand, such differences are also evident in most other economic sectors, where they are generally accepted in the short run and become the subject of international harmonisation efforts in the longer term.

Environmental risks and their consequences are perceived very differently from country to country, even when some of the consequences have potential impacts far beyond national borders. In some countries, the general public broadly supports nuclear power or at least accepts it; in others, it is strictly prohibited. Binding international security standards are in place, but various aspects in the approval process, in the management of spent fuel and in the management of long-term waste disposal vary from country to country. All in all, this creates differences that influence the costs of nuclear power from one country to another. Many aspects related to nuclear power have a broad policy interest within and across countries. However, as long as conditions for nuclear power are transparent, related policy initiatives are unlikely to undermine the functioning of liberalised electricity markets.

Support for R&D and differing legal conditions for different technologies both affect competitiveness in open electricity markets, but only indirectly and on a level comparable with many other economic sectors.

Liberalisation significantly alters decision-making processes, transferring them from the centralised and planned structures within vertically integrated systems to the decentralised decisions made by individual market players, based on direction of market prices. Direct financial subsidies to specific technologies, such as renewables and nuclear power, perhaps in specific volumes and perhaps even on specific locations, fundamentally undermine the decentralised decision-making process in liberalised markets. Alternative policy options must be applied to internalise external environmental costs while at the same time maintaining the decentralised decision process. When taxes are not viable, the least distortive policy option is the cap-and-trade market mechanism. As with the tax option, cap-and-trade systems still focus on the actual harmful activity but it does not stipulate or dictate the solutions as direct subsidy schemes do.

For example, a cap-and-trade system is stipulated for greenhouse gases within the flexible mechanisms of the Kyoto Protocol. The European Union implemented a cap-and-trade system for CO_2 emissions. During the first half of 2005, trade of CO_2 emission permits increased dramatically. Liquidity and price of the permit market reached a level at which this market is commonly accepted as one of the explanatory factors for price increases in most European electricity markets. National governments determine the conditions for entering the European CO_2 emission permit market for each individual generator. These conditions differ significantly and leave the European electricity market severely distorted. However, the marked-based mechanism itself has already demonstrated its ability to create the intended and anticipated incentives. The first experiences with CO_2 emissions trading are explored in a recent IEA publication.[97]

Most IEA member countries support renewable energy. It, too, can be supported in a cap-and-trade system. Several member countries, including the United Kingdom, Australia and Sweden, as well as several U.S. states, introduced systems based on quotas for renewable energy and tradable renewable certificates, known as renewable portfolio standards (RPS). Policy intervention determines volumes but investors decide on technology, time and place. If the justification for supporting renewable technologies accrues to the fact that these technologies does not emit greenhouse gases, full implementation of a cap-and-trade system for greenhouse gases will make an RPS obsolete. Hence, if the justification is based on the assumption that renewable energy is a cost-effective way to reduce emissions, an increasing CO_2 emission price will eventually reduce the price of renewable certificates to zero. RPS systems will only be robust, cost-effective policy options with minimal distortion of electricity markets if they are international and consistent with efficient operation of open electricity markets. A purely national RPS will be vulnerable to political changes and will lose the tremendous efficiency potential of international trade. Most renewable resources are highly locational and, thus, significantly more cost-effective if deployed where resources are abundant or diverse, *e.g.* wind turbines where it is windy, bio-mass plants where there is a lot of cheaply available biomass and photovoltaics where it is sunny.

97. IEA, 2005d.

Many renewable energy technologies operate in smaller unit sizes than conventional generation technologies and tend to be more distributed and connected to the grid on lower voltages. Most newer technologies, such as wind power, are also highly intermittent. In the traditional vertically integrated and regulated electricity industry there is a natural focus on large conventional generation plants and systems are developed and operated to meet both the needs and constraints of these technologies. Moreover, the costs are passed on directly to consumers, so there are no built in incentives to enhance transparency. With such a focus it becomes difficult to fully assess the costs and benefits of distributed and intermittent renewable technologies. But without an overview of costs and benefits, it is impossible for policy makers to fully assess the merits of renewable technologies, including their cost-effectiveness in terms of mitigating greenhouse gas emissions. Liberalised electricity markets change the framework for distributed and intermittent renewable technologies on several accounts. Cost-reflective pricing and competition determine the real costs of managing intermittency. The larger focus on risks, coupled with the value of financial and operational flexibility, in liberalised markets sheds new light on the merits of small, distributed generation resources. Cross-border trade makes it possible to use local renewable resources much more effectively, for example, by exploiting a diverse resource base in wind.

Some of the costs and benefits of renewable energies accrue to the impact that distributed and intermittent technologies have on network development and operation. On one hand, high shares of intermittent wind necessitate network extensions; on the other, wind turbines and other distributed resources connected to local networks may save some costs to grid losses and some investment in transmission networks. Creating incentives for network tariffs that reflect these costs and benefits is a challenge for effective economic regulation of networks. Economic impacts of integrating wind power into electricity grids are discussed in greater detail in the annex of a recent IEA/NEA study on projected costs of generating electricity.[98]

There is intense debate over the merits of renewable energy technologies, particularly wind power which seems to be one of the more cost-effective options in some circumstances. For those countries with the highest shares of wind power as a percentage of consumption (*e.g.* Denmark, Germany and Spain) there is concern for the security of national and neighbouring electricity systems. In addition, there is increasing concern about the cost-effectiveness of wind power when all costs of managing intermittency are

98. IEA/NEA, 2005

taken into account. In liberalised electricity markets, all of the issues associated with management of intermittent renewable technologies are more - or even fully - transparent. Competitive locational marginal pricing of energy and competitive pricing of reserves and ancillary services also make the costs and benefits of intermittency and geographically diverse generation more transparent and the total costs possibly lower.

Box 6 . Managing intermittent wind power in the liberalised Danish electricity market

Denmark is divided into two electricity systems that are not interconnected. The eastern system is electrically synchronised with the Nordic system; the western with the continental European UCTE system. By the end of 2004, some 3 122 MW of wind power capacity was installed in Denmark. During that year, wind turbines produced 6.6 TWh, corresponding to 19% of the consumption in a year with average energy content in the wind at 90% of the normal value. Approximately 2 379 MW of the total installed wind power capacity was located in western Denmark, where turbines generated 4.9 TWh, or 23% of consumption for that region.

High shares of wind power in the Danish system are at the heart of a serious debate about costs, system security, industrial policy and impact on landscape that has been ongoing throughout the past decade of wind power development. Many issues regarding the impacts on the Nordic electricity market still remain unresolved. Until recently, Denmark subsidised wind power through feed-in tariffs given for every MWh produced from wind power; the tariff was reduced in 2003 to a level that has literally halted further development. Instead, most new wind power development is now driven through politically decided tenders for off-shore wind parks and politically decided refurbishment schemes. Such an approach deviates fundamentally from the principles of decentralised decision making in liberalised markets. For example, Sweden implemented a cap-and-trade system to support wind power and Sweden and Norway decided to develop a common cap-and-trade system. Some harmonisation issues related to policy - and its impact on broad investor confidence - in the Nordic market remain open. However, there is no doubt that the Nordic electricity market has been very important for the development of wind power in Denmark. This is evident when one compares data on hourly wind power production in western Denmark and the hourly Nord Pool day-ahead spot price for December 2003 (Figure 18).

Figure 18

Wind power production and Nord Pool day-ahead prices in western Denmark, December 2003

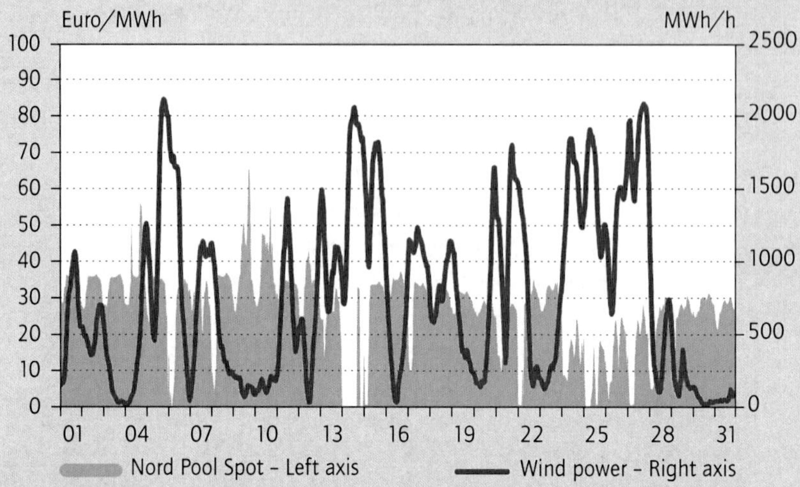

Many factors influence electricity prices in western Denmark; Figure 18 shows clearly that wind power is one of them. The price tends to drop when wind power production is high and rise to prices in neighbouring markets when production is low. Transparent spot and balancing prices have significantly improved understanding of costs and benefits of wind power. Moreover, the liberalised market has directed trade across the Nordic region in a way that is sufficiently dynamic to enable integration of Danish wind power across a larger area – even while the direct costs of Danish renewables policy remain with Danish electricity consumers. Figure 19 shows the impact this price development has on trade, as well as hourly wind power production and hourly aggregate exchange with Norway and Sweden.

Exchange between the Nordic countries is entirely determined by the price of electricity for every hour. Figure 19 shows that the price directs wind power production to Norway and Sweden when the level is high. This is a sensible outcome, particularly considering that Norway has large hydro reservoirs, which are excellent for storing electricity. Danish electricity consumers pay the direct costs of Danish renewables policy, but the costs are minimised through international trade.

Figure 19

Wind power production in western Denmark and trade with Nordic neighbours

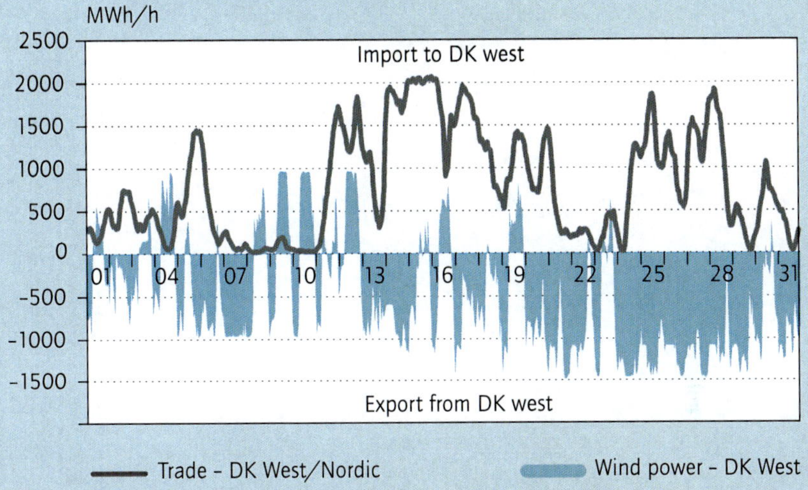

ANNEX 1: BRITISH ELECTRICITY TRADING AND TRANSMISSION ARRANGEMENTS

Privatisation was the main driver for the launch of the liberalisation process in the late 1980s. The process was triggered by the publication (February 1988) of a government white paper, *Privatising Electricity*. The white paper outlined a number of principles including: competition is the best guarantor of customers' interests; regulation should be designed to promote competition, oversee prices, and protect the customers' interests in areas where natural monopolies will remain; and security and safety of supply must be maintained. The white paper stipulated an eight-year transition period. Prior to the publication of the white paper, the telecommunications and gas businesses had been privatised and a reform of the state-owned coal business had been initiated.

The white paper was followed up with a legal framework enabling the effective launch of liberalisation; the Electricity Act 1989. The most critical initial parts of this framework were the establishment of an independent regulator and a restructuring of the sector. Under the Electricity Act 1989, the post of Director General of Electricity Supply (DGES) was created to regulate the natural monopoly businesses. The government established the Office of Electricity Regulation (OFFER) as an independent body to be headed by the DGES. Its role was to promote competition and regulate the network businesses. OFFER established a new regime for network regulation based on a cap of prices, with a focus on cost-cutting. It was the Retail Price Index (RPI-X) regime, which limited the development of prices by a cap consisting of the increase of the RPI minus a specified demand for improvement in efficiency. This model was to be copied by many regulators in the following years.

Prior to the Electricity Act 1989, the entire electricity industry in England & Wales was state owned. Generation and transmission were managed by the Central Electric Generating Board (CEGB) and distribution was managed by 12 area electricity boards. Under the Electricity Act 1989, the entire sector was reorganised, corporatised and eventually privatised.

CEGB was split into four companies. All transmission assets and responsibility for system and market operation was transferred to National Grid Company (NGC). All generation assets were divided between National Power (40 conventional power stations with 30 GW capacity), Power Gen (23 conventional power stations with 20 GW capacity) and Nuclear Electric

(12 nuclear power stations with 8 GW capacity). Initially it was intended to place the nuclear assets with National Power but these plans were abandoned at a late stage on financial advice saying that the nuclear assets were not saleable at a reasonable price. The 12 area electricity boards were corporatised into 12 regional electricity companies (RECs). Ownership of NGC was transferred to the 12 RECs. The vesting of CEGB and the 12 area electricity boards took effect on 31 March 1990. A series of three-year contracts were signed. National Power and Power Gen signed contracts for purchase of coal from the still state-owned British Coal at a price above world market prices. These generators held contracts with the 12 RECs enabling a pass through of the extra cost to captive customers. On 1 April 1990, retail competition opened to the 5 000 consumers with a load higher than 1 MW and the Pool commenced operation.

In Scotland, the North of Scotland Hydro-Electric Board was also restructured under the Electricity Act 1989 into Scottish Hydro-Electric and Scottish Power. Both were privatised as vertically integrated regulated utilities in June 1991.

The first round of privatisation happened in December 1990 with the 12 RECs. The government still kept a golden share that hindered mergers and takeovers. These golden shares lapsed in April 1995 and immediately triggered the first trading activities with REC assets. In March 1991, the first 60% of National Power and Power Gen were privatised and the remainder were privatised in March 1995. In December 1995, NGC was sold by the RECs through a flotation of NGC shares on the stock market. Initially, the pumped storage capacity of CEGB had been allocated to NGC but this was sold when NGC shares were floated. The final round of privatisation happened in July 1996. The nuclear power stations in Nuclear Electric were transferred to British Energy (the seven modern reactors, excluding the seven old Magnox reactors) and British Nuclear Fuels Ltd. (the seven Magnox reactors). Also two reactors within Scottish Electric were transferred to British Energy. British Energy was then privatised.

The three-year vesting contracts and the five-year golden shares in RECs gave some time to develop the operation and functioning of the market. When the three years ran out, prices started to increase and the DGES raised concern over the development of competition in the market. Several initiatives were taken that were to have significant influence on the development of competition and the structure of the industry. Competition and divestiture were the only sources for increased competition. In anticipation of the importance of new entry by independent power producers (IPPs), it was accepted that RECs could enter into long-term power purchase agreements from IPPs and IPPs could sign long-term gas contracts. This ignited the "dash for gas" that was to add significant capacity of new combined cycle gas turbines (CCGTs) and decrease the market power of the incumbent generators. The dash for gas more than halved the size of the remaining deep coal mining industry. When margins between fuel costs and Pool prices did start to increase after 1993, the largest incumbent, National Power, agreed to accept divestiture of 6 000 MW generation capacity within two years under the threat of being referred to the Monopolies and Mergers Commission, the competition authorities. A price cap was also introduced during the two years. When the golden shares of the RECs lapsed, these were bought by other regulated UK utilities and two US utilities. Power Gen and National Power also gave bids for two RECs, driven by the desire to vertically integrate with retail supply businesses. The bids were referred to the Monopolies and Mergers Commission, and in the end blocked by Department of Trade and Industry. The response from the two incumbents was to accept the divestiture of an additional 4 000 MW generation capacity each, in exchange for the admittance to acquire RECs. This happened in November 1998. Several large foreign utilities have entered the British market and some have left again. National Power, with its reduced generation capacity but with its REC arm, was bought by RWE. Power Gen was bought by German E.ON. French EDF bought two large RECs including London Electric.

An increasing reliance of gas and increasing problems for the British coal industry losing market shares to gas raised concerns with the new Labour government in 1997. In December 1997, the government imposed a moratorium on building new gas-fired power stations, which halted development. The Department of Trade and Industry estimates that the moratorium delayed the building of 5 200 MW of new gas-fired generation capacity.

Retail contestability was extended in steps. The barrier was lowered from 1 MW to 100 kW in April 1994, thereby increasing the number of contestable consumers to 45 000. The remaining 22 million consumers were given access to switch supplier between September 1998 and June 1999.

This ended the process that the Electricity Act 1989 had triggered 10 years earlier. But already before that, the first considerations to reform the market which had been implemented were starting to emerge. In October 1997, the Minister for Science, Energy and Technology asked the DGES to review the electricity trading arrangements in the pool. The merits of gas compared to coal in the pricing principles of the Pool were one of the reasons for initiating the Pool review. OFFER published its Pool Review in July 1998, which included a far-reaching critique of the existing mandatory pool system and gave recommendations for new electricity trading arrangements (NETA) based on a voluntary approach. The government accepted the recommendations in a white paper with conclusions from the review in October 1998, which triggered the reform of the Pool. NETA replaced the Pool on 27 March 2001. In August 2000, NGC established a separate company to manage the new Balance and Settlement Code, the rule book for NETA that all market participants must sign. The new company, ELEXON, a subsidiary to National Grid Company operates and settles the balancing market in NETA.

A new Utilities Act 2000 was passed in July 2000. The main points were: replacement of separate gas and electricity regulators with one regulator, the Office of Gas and Electricity Markets (OFGEM); legal separation of retail supply and distribution; and enabling the implementation of NETA. The Utilities Act 2000 forced the legal unbundling of the two vertically integrated Scottish utilities and discussions about the extension of NETA into Scotland commenced.

After a decade with focus on cost-cutting in distribution networks concerns were raised for the quality of service. In 2003, OFGEM introduced a new aspect in the RPI-X regulation of prices, specifically targeted at mimicking incentives for efficient levels of quality. By measuring the quality of service in terms of the number of interruptions of supply, the duration of these interruptions and the information service provided in connection with these interruptions, failures to perform according to acceptable standards could lead to a reduction of prices of up to 1.75%.

Scotland was integrated into NETA in April 2005, which is hereafter referred to as the British Electricity Trading and Transmission Arrangements (BETTA). NGC was appointed as market and system operator for BETTA with ELEXON in the role as market operator.

Framework for Governance

The Utilities Act 2000 establishes the framework for the British liberalised market. It requires independence of transmission ownership, system and market operation from the competitive businesses in the sector. This independence is through ownership unbundling. It requires legal independence of distribution networks from generation and retail supply. The Utilities Act gives British electricity consumers the freedom to choose retail supplier. It gives the authority to regulate the electricity industry to the Gas and Electricity Market Authority (GEMA) with the regulatory body Office of Gas and Electricity Markets (OFGEM).

OFGEM's role is to regulate the electricity and gas industries in protecting consumers by promoting competition where this is possible and through regulation where this is necessary. More specifically, OFGEM focuses on making electricity and gas markets work effectively and intelligently regulating monopoly businesses. OFGEM is governed by GEMA, a body appointed by the secretary of state.

National Grid Company has been licensed by the GEMA to operate the British electricity system and, through its wholly owned but independently managed subsidiary ELEXON, to operate the balancing market. OFGEM regulates the performance of National Grid Company and approves the grid code that sets the rules for interaction with the British electricity system. The balancing mechanism is the only official market place in BETTA. It is functioning in accordance with the Balancing and Settlement Code (BSC). One of the reasons for abolishing the Pool was that the governing structures made it very complicated and tedious to make any changes to the rules. As a consequence, a very strict and clear procedure for managing changes of the BSC has been established, in which the GEMA has a stronger influence. In accordance with these modification procedures, ELEXON established a BSC panel to ensure an impartial enforcement of the code and to give impartial advice on any proposed modifications. The Panel consists of representatives appointed by GEMA (appoints the chairman), industry, EnergyWatch and National Grid Company. EnergyWatch is the independent watchdog for electricity and gas consumers. ELEXON advises the BSC Panel. Any modifications to the BSC are subject to approval by GEMA, which must justify its rulings transparently.

OFGEM has the responsibility to give licenses to distribution network operators, who are then subject to OFGEM RPI-X and quality of service regulations. In connection with quality of service regulation, annual reports are published on the quality of service.

OFGEM also has the responsibility to monitor and secure the supply of gas and electricity. In its efforts to fulfil this responsibility, OFGEM publishes bi-annual reports on the security of supply of gas and electricity.

Basic Market Design Features

The initial trading arrangement in the Pool was inspired by the principles applied in the calculation of dispatch by the former CEGB and the dispatch in the Pool was calculated with the same software, GOAL. Generators gave bids and specified start-up costs and other technical constraints. Bids were ordered in ascending order and the software calculated the dispatch that would meet forecast demand, subject to transmission constraints. The marginal bid set the system marginal price to be paid to all dispatched generators. Grid losses and congestion management (locational aspects) were not reflected in the pricing; they were only taken into consideration in the actual dispatch. Demand was forecast by National Grid Company, so it was a one-sided market assuming no demand response to prices. Bids and prices were calculated for every half hour of the following day. It was a day-ahead trading mechanism, in which bids had to be submitted by 10:00 a.m. one day ahead, and dispatch was calculated during the afternoon before 5:00 p.m.

The system marginal price was only one element in the price paid by consumers and the remuneration received by generators. The most significant addition was the capacity payment paid to generators for keeping generation capacity available to the market. It was awarded as a price-per-available-generation capacity but the price was calculated dynamically from an assessment of the reserve margin, the probability of losing load and an assumed cost of the lost load for every half hour. Hence, the price increased with decreased reserve margins, also if this was the result of withholding generation capacity. This system for capacity payment was a sincere effort to reflect to true value of capacity for every half hour, but it also proved to be prone to manipulation in a market with dominating generators, as was the case in the England & Wales market in the mid-1990s. In addition, an uplift was charged from consumers to pay for transmission losses, ancillary services, reserves and other necessary services to operate the system.

All incumbent generators possessed the GOAL software, which gave them considerable power to maximise profit through their bids, rather than bidding the true competitive cost parameters of their plants. Apart from the problems with market power abuse, there were also a range of other critique points. It was only a one-sided market, which made demand participation difficult. The rules for trading in the Pool were set in the Pooling and Settlement Agreement, a multi-lateral agreement signed by all market participants. It was very difficult to make the necessary changes to this agreement, allowing the market to develop. At the initiation of the Pool review in 1998, a liquid financial market also failed to develop, mainly because the underlying market was uncompetitive.

The approach taken in NETA, later extended to BETTA, was fundamentally different. Instead of basing the trade on obligatory trade in the Pool, trading was to be entirely based on voluntary bi-lateral trade divided into four market segments. The intention was that a forward market for standardised financial contracts should develop for the long run, covering periods up to several years in advance. A market for bilateral contracts traded in the OTC market should develop for the medium and long run. A short-term OTC market for bi-lateral contracts should develop for the last 24 hours before gate closure. All these markets were entirely voluntary and were intended to develop automatically, driven by the need from market participants. The only market that market participants are forced to link into is the balancing market, operated by National Grid Company and ELEXON. Gate closure is one hour before real-time operation. All registered market participants are obliged to submit schedules (Final Physical Notification) before gate closure and will be held financially responsible for any deviations from the schedule. Schedules are submitted both for consumption and generation, making it a real two-sided market. Actual trade in the balancing mechanism is still voluntary. All market players are free to manage unexpected deviations in generation and demand.

The balancing mechanism is an adjustment market. Market participants can submit bids to the balancing markets with prices and volumes that they are willing to increase and/or decrease generation and/or demand with. NGC calls on the cheapest bids to balance the overall system physically. The purchase of balancing services deviates fundamentally with all other similar markets on one critical point. The price paid to those that are called on to deliver balancing services are pay-as-bid or discriminatory pricing, meaning that the prices bid by each individual market player is also what he will receive if called on. This is fundamentally different from the marginal pricing principle used in all other electricity markets. There has been extensive research into the merits of these two pricing principles. The philosophy with pay-as-bid is that only those generators (or consumers) that are able to deliver at lower cost than the most expensive will receive this lower price. Those that buy the service will benefit from this technological comparative advantage. Average prices are thereby thought to become lower. The problem is that those with the comparative advantage are not necessarily willing to pass this advantage on to rate payers. To avoid that, they may want to bid in at prices that are as close to, but slightly below the most expensive resources needed. This puts pressure on them to focus more on the costs of their competitors, rather than their own costs. As a result, average prices may not be lower, and may even be higher as a consequence of speculation. Research has been inconclusive but has illustrated that the alleged merits of pay-as-bid markets are overstated. Market participants do not disclose their true cost structure when bidding into pay-as-bid or discriminatory auctions.

Those that cause the overall system imbalance between scheduled and actual operation are financially responsible for the costs of those imbalances. The charges for these individual imbalances have changed since the implementation of NETA. In the current pricing principle, those that have imbalances in the same direction as the total system imbalance are charged with the weighted average price of those that were called on by NGC to physically balance the system in each half-hour period. Those that have imbalances in the other direction, and who thereby have helped the overall system imbalance by chance, are charged with a spot reference price. The spot reference price is the day-ahead spot price from the privately owned UK Power Exchange (UKPX). Liquidity in UKPX in terms of turnover as a percentage of total British demand is very low, which could be seen as a threat to the role of this spot price to serve as a reference. Another problem that has been raised with the issue of such a dual pricing principle is that it creates an incentive to pool imbalances. The dual pricing principle generates a profit for NGC. This residual cash flow is paid back to market participants, corresponding to their proportion of total demand and generation. A merger of companies lowers imbalance charges without necessarily improving forecasting or lowering the real costs of imbalances in the system. Another related issue of concern is that it is specifically beneficial to pool generation and load to enable self-management of imbalances within a company. This lowers liquidity in the "official" balancing market and may increase total system costs of imbalances in the end. More expensive resources within a company may be used to manage imbalances instead of benefiting from other cheaper resources owned by other generators.

Congestions within the British grid are not priced in the balancing mechanism. There is one price for the entire BETTA. On the other hand, there are locational signals in the transmission network tariffs for connecting to and using the transmission grid.

The remaining three markets that were pictured to develop automatically have so far not materialised. An illiquid spot market has developed. Initially, several projects to establish day-ahead spot exchanges were launched, but there is currently only the UKPX with very low traded spot volumes. There is not much transparency in the OTC market, but liquidity is probably relatively low. There are no standardised forward contracts traded on exchanges. Prices in NETA have been deemed close to competitive levels according to models of competitive pricing and physical assets have been traded extensively, so with that measure the British market has been liquid.

With a market philosophy based on voluntary trading, transparency has been a matter for the voluntary exchanges. As these have not materialised so far, market transparency has been low. Prices and traded volumes in the balancing mechanism are easily available from the homepage of ELEXON. There are, however, no requirements to make changes in the status of generation plants immediately available to the market. National Grid Company submits annual reports to the GEMA with a seven-year statement of forecast development of peak load, installed generation capacity and transmission capacity. The statements are easily available on the NGC Web site and provide an easy access to extensive fundamental data.

Market Structure

In 2003, electricity consumption in the United Kingdom was 348 TWh of which 313 TWh was in England & Wales, 35 TWh in Scotland and 8 TWh in Northern Ireland. To serve the 26 million residential and 2.5 million commercial consumers, some 376 TWh was generated in 2003 and 2 TWh was imported. The remaining 30 TWh were transmission and distribution losses. Approximately 22% was generated on nuclear plants, 37% was generated with gas, 35% was generated with coal, 1.5 % was generated with hydro and the remaining was generated with other sources.

Demand in United Kingdom peaked at 61 GW in the winter 2004/05 of which a little more than 1 GW was in Northern Ireland, outside the BETTA area. To meet this peak demand, there was an installed capacity of 75 GW of which 16% was nuclear, 37% was coal, 33% was CCGTs, 5% was hydro and the remaining was from other technologies. In addition, there is an interconnector between France and Britain with a capacity of 2 000 MW; Northern Ireland and Scotland are interconnected with 500 MW; Northern Ireland and the Irish Republic is interconnected with 600 MW, and Scotland and England are interconnected with 2 200 MW to keep BETTA united.

By mid-May 2004, the three largest generation companies registered by Department of Trade and Industry statistics were British Energy with 11.5 GW of nuclear capacity (16% of total installed capacity), RWE Innogy with 8 GW coal, oil and CCGT capacity (11% of total installed capacity) and E.ON UK with 7.6 GW of installed coal, oil and CCGT capacity (10% of total installed capacity). Other large companies include Scottish Power, Scottish and Southern Energy, EDF Energy and American Electric Power, all with 5% to 6% shares of total installed capacity. An additional 37 companies were operating generation plants in United Kingdom.

While market shares in generation are very widespread without any dominating players, the concentration in the retail market is higher with six large companies. By the end of 2003, the largest retail suppliers were British

Gas (24 % of the market), Powergen (21% of the market), npower (15% of the market), EDF Energy (14% of the market), Scottish and Southern Energy (14% of the market) and Scottish Power (11% of the market). The remaining percentage was shared between other small retail suppliers. Comparing to the original distribution companies linked with the 12 area electricity boards and the two Scottish boards, there has been a significant concentration. Several of the retail suppliers are also major generators. British Gas has benefited from marketing electricity and gas jointly.

Further Reading

Background and development of the electricity supply industry is described in IEA country reviews. The joint reviews of the United Kingdom since late 1980s constitute a comprehensive description of the development and of governing structures. The Web sites of Department of Trade and Industry (www.dti.gov.uk), OFGEM (www.ofgem.gov.uk), National Grid (www.nationalgrid.com) and ELEXON (www.elexon.co.uk) include many descriptions, information, statistics and rules. Substantial research has been undertaken in analysing the functioning of the Pool, the functioning of NETA, the development of competition and the development of the structure of the sector. Hunt (2002), Evans & Green (2003) and Newbury (2005) are some examples.

ANNEX 2: THE NORDIC ELECTRICITY MARKET

The first step towards a competitive internal Nordic electricity market was taken with the approval by the Norwegian parliament of a new electricity reform act in June 1990. The act came into force on 1 January 1991, and introduced regulated third-party access, freedom of choice of retailer for all electricity consumers and unbundling of the transmission grid and system operation. On 1 January 1992, the state owned utility Statkraft was broken up into two independent state owned companies, Statkraft with all generation assets and retail contracts from the former company, and Statnett with all transmission assets (85% of total Norwegian transmission assets) and responsibility for system operation.

In 1993, the Nordic power exchange was established as an independent company under the name Statnett Market AS. It established price quotation on a day-ahead basis and it established the world's first exchange-based trade with futures contracts in July 1993. An important milestone for the development of the retail market was the reduction of a standard fee for retail switching from NOK 4 000 (app. EUR 500) to NOK 200 (app. EUR 25) in 1995. From 1997, there were no fees for switching and from 1998 every consumer could switch supplier with one week's notice.

A second decisive process towards an internal Nordic market was initiated with the approval of a new electricity reform act by the Swedish parliament in May 1992, which established the framework for initiating competition on wholesale level and set a plan for additional steps to create competition. It had been preceded by steps towards the introduction of competition already in 1991. With effect from 1 January 1992, the State Power Board was transformed into a state-owned limited liability company, Vattenfall AB, and the transmission system, including cross-border interconnectors, were separated out into a state entity, Affärsvärket Svenska Kraftnät. Svenska Kraftnät operated the transmission system and was also instructed to promote competition. The most important reforms were laid out in the Electricity Act passed by Parliament in October 1995, with effect from 1 January 1996. The act required legal separation of generation and network operation. It gave all consumers the freedom to choose supplier. In the wake of the act, it was decided to establish a common Norwegian and Swedish electricity exchange under the name Nord Pool. It was established with day-ahead price quotation from 1 January 1996 and extended with trading in financial futures contracts in 1997. Nord Pool ASA was established as a Norwegian company with headquarters in Norway and owned by the two transmission system operators, Statnett and Svenska Kraftnät with 50% shares each.

The liberalisation process in Norway and Sweden ran almost in parallel, but actual reform to introduce competition proceeded at a faster pace in Norway. One of the important drivers for regulatory reform in Sweden was the relatively severe economic recession at the time. Consequently, there was a high focus on reducing electricity costs, which were critical considering the high share of heavy, energy-intensive industry in Sweden.

The next country to be integrated into the Nordic market was Finland, a process initiated with the approval of the new Electricity Reform Act that came into force on 1 June 1995. It gave regulated third-party access to the grid to everybody except consumers with a power requirement below 500 kW, a threshold that was removed on 1 January 1997. An independent regulator, the Electricity Market Authority, was established with the Act. A Finnish TSO, Fingrid Oyj, was established by the government in 1996 and commenced operation on 1 September 1997. Fingrid was formed after the unbundling of the transmission grids from the two large vertically integrated utilities, IVO (owned by Fortum) and PVO. Fortum was 75% state owned, a share now reduced to 59%. The ownership of Fingrid is distributed between the utilities Fortum (25% shares, 33% votes) and PVO (25% shares, 33% votes), the Finnish State (12% shares, 16% votes) and the remainder with institutional investors.

The Finnish electricity act did not establish a framework for the development of trading arrangements. Two electricity exchanges were established through private commercial initiatives in 1995, EL-EX and Voimatori Oy. Liquidity never reached satisfying levels. The two exchanges merged into EL-EX in 1996. When liquidity still did not materialise Fingrid bought all shares in 1998 and from 1 June 1998 EL-EX became Nord Pool's representation in Finland. Finland, thereby, effectively entered into the internal Nordic market with a day-ahead price quotation at Nord Pool.

Denmark entered the Nordic market in two steps, thereby completing the picture.[99] Denmark has two physically independent electricity systems, one being synchronised with UCTE and one being synchronised with the Nordel system. The two systems are not physically connected. To a great extent, liberalisation in Denmark was driven by the development in the other Nordic countries, where Danish utilities started to take part in the trade, and by the first EU market directive. In anticipation of the expected development, Danish

99. Iceland is also a Nordic country and a member of Nordel. Iceland has taken some steps to reform their electricity sector, but it is not interconnected with the other Nordic countries. What is referred to as the Nordic market in this book is excluding the Nordic country Iceland.

utilities started to adapt to the market development ahead of the legislation. The dominant utility in the western part, Elsam, re-organised with the establishment of a separate business entity for transmission ownership and system operation, Elsam System, from 1 January 1997. Elsam System was turned into a separate company, independent of Elsam, with effect from 1 January 1998 under the name Eltra. An amendment of the Danish electricity act in May 1996 transposed the EU market directive into Danish law by giving large consumers (more than 100 GWh per year) access to the grid from 1 January 1998. However, the act did not require unbundling and did not specify the framework for grid access. This only happened with the approval of the Electricity Supply Act on 2 June 1999 which established a framework for competition through regulated third-party access, legal unbundling of networks and the establishment of a regulator independent of government. In the western part of Denmark, market rules very similar to those found in the other Nordic countries had already been developed through 1998 and the spring of 1999. Thus, a western Danish market was launched on 1 July 1999 with a price quotation on Nord Pool, immediately after amendment of the electricity act. Eastern Denmark followed with the formation of an independent TSO, Elkraft System, on 1 January 2000 and Nord Pool price quotation on 1 October 2000. The Danish electricity sector was more or less fully owned by municipalities and co-operatives. From 1 January 2005, the Danish state bought the shares of the two TSOs and a joint, state-owned TSO was formed under the name Energinet.dk, also including system operation in gas.

Danish electricity consumers have gained access to the grid in steps with consumers above 10 GWh/yr getting access from 1 April 2000. On 1 January 2001, the threshold was lowered to 1 GWh and it was removed altogether as of 1 January 2003.

A final important milestone in the development of an integrated Nordic market is marked by the re-organisation of Nord Pool in 2002. By that time, Nord Pool conducted four key activities: the day-ahead spot exchange, financial trading, clearing and consultancy. Nord Pool was still owned by Statnett and Svenska Kraftnät. In early 2002, the day-ahead spot exchange was transformed into a separate company. From 1 July 2002, the other Nordic TSOs joined Nord Pool Spot as co-owners with 20% shares to each country and the last 20% still remaining in the hands of Nord Pool Holding.

The Nordic electricity market developed constantly throughout the period. Market design has been refined and harmonisation for further integration has taken place in steps, driven by agreements at regular meetings between Nordic energy ministers. Nordic energy authorities have a permanent working group to facilitate the necessary co-ordination between authorities and TSOs. Nordic TSOs co-operate through the association Nordel, which is the framework for permanent committees and numerous working groups that practically address a whole range of issues related to operation, system planning and market design.

Framework for Governance

In all Nordic countries, electricity market liberalisation is founded on a relatively detailed legislative framework. Legislation specifies the freedom of choice for consumers, regulated third-party access and legal unbundling of network activities. Finnish and Danish electricity reform acts also established the framework for the formation of regulators independent of government (regulators were already in place in Norway and Sweden). In Denmark, the Energy Agency, a government body, also has some of the roles that regulators traditionally have, including issuing licences for network companies.

The roles and responsibilities of TSOs are specified in great detail in the legislations of each Nordic country. The common responsibilities are to ensure, maintain or enhance: operational security of the system; the momentary balance between demand and supply; adequacy of the transmission system in the long term; and efficient functioning of the electricity market.

All TSOs co-operate closely with stake holders in the industry. Rules and principles that guide market design and information flow are conceived in various ways, with slight deviations from country to country. TSOs are involved, in one way or another, in most interactions that occur in the market and often play a central role in rule making and market design. Regulators monitor and often contribute to the process; other stakeholders are given a chance to react.

Regulators oversee network tariffs and the performance of network companies, including TSOs. The Norwegian regulator, Water Resources and Energy Directorate (NVE), regulates network tariffs according to a cap system that allows revenues to increase with CPI, minus an annual requirement of efficiency improvement. NVE is a subordinated ministerial agency with independence in its day-to-day dealings. The Ministry of Petroleum and Energy is, however, the court of appeal for NVE rulings. The Swedish regulator, the Swedish Energy Agency, oversees the implementation of the electricity act, including ensuring that network tariffs are reasonable. It is governed by its board, which is appointed by the government. In Finland, the Energy Market Authority regulates networks by monitoring tariffs and the implementation of

Finnish legislation. It is the sole authority to oversee the implementation of the electricity act and it is independent from the Ministry of Trade and Industry. In Denmark, the Danish Energy Regulatory Authority oversees that network tariffs are in accordance with legislation and the directives established by the Danish Energy Agency. It is governed by a board, appointed by the relevant minister for a fixed term.

Competition is regulated and monitored by competition authorities in all Nordic countries.

Basic Market Design Features

A first crucial building block in the design of the internal Nordic market is the postage-stamp grid tariffs. Across the entire Nordic system, there is a tariff for feeding electricity into the grid and a tariff for taking electricity off the grid. Initially, there were examples of specific border tariffs but the last element of that was removed when Sweden abolished its border tariff towards Denmark (March 2002). Tariffs are different from country to country – and even from region to region – but these differences mainly reflect real costs. There is no element in the network tariffs that unduly distorts trade within the Nordic region.

The day-ahead spot market organised through Nord Pool is the cornerstone of the internal Nordic electricity market. Bids and offers must be submitted to Nord Pool by 12:00 noon for the following day. Bids and offers are put in merit order: the marginal bids and offers that determine the balance between supply and demand sets the price for the entire market. Unit commitments or other considerations regarding fixed costs are not taken into account in the market clearing but market players have various opportunities to submit block-bids. Block-bids enable generators to make a bid conditional of being dispatched for a block of hours instead of only one. Nordic TSOs give Nord Pool Spot a monopoly to use all available transmission capacity that interconnects the defined areas or zones in the Nordic market. Currently there are three zones in Norway, but they can change if Statnett sees frequent congestions in other places. Denmark has one zone for each of their two separate systems. Sweden and Finland constitute one zone each. Nord Pool calculates market prices for each zone, taking all bids and offers in each zone and interconnection capacity between the zones into account. The resulting zonal prices clear the market. All network companies are responsible for assessing and purchasing electricity resulting from grid losses. Hence, grid losses are reflected in the zonal prices through normal demand bids in the spot market.

Nord Pool also calculates a system price assuming that there are no constraints in the entire Nordic transmission system. This is purely a reference used in the financial market and does not necessarily correspond with the exact prices faced by any market players. So far, market outcomes often show Norwegian zonal prices almost identical with the system price. Prices in the entire Nordic market are regularly identical and equal the system price. The western Danish zone is the most clearly congested area as it constitutes the gate between the hydro-based Nordic electricity system and the more traditional thermal electricity system in continental Europe.

Initially, some interconnection capacity was reserved for old, long-term contracts. The last of these reservations was removed in 2000 when Stattkraft, Elsam and Preussen Electra (latter to merge into E.ON) abolished contracts that blocked the Danish-Norwegian interconnection. The parties were financially compensated by Eltra, the western Danish TSO.

The transmission capacity made available to Nord Pool, as announced during the morning before day-ahead bids and offers are submitted, is guaranteed by the TSOs. This implies that the transmission right is firm. It is the responsibility of the TSOs to re-dispatch if the announced transmission capacities turns out to be unavailable at the moment of operation. The costs of such a circumstance are a part of transmission network tariffs and are thereby paid by all consumers and generators. Conversely, available transmission capacity is also a source to collect congestion rents, which will be used to lower tariffs if they are not used for financing new interconectors.

Nord Pool operates a Web site on which prices, volumes and other fundamental market data are easily available. All market participants that wish to trade on Nord Pool must sign a user contract that outlines a wide range of obligations and responsibilities defining the interaction between market participants and the market. These obligations include a requirement to immediately disclose to Nord Pool any changes in generation and transmission facilities larger than 50 MW. Such changes are immediately posted on Nord Pool's information service as a market message. Market messages include planned revisions and urgent messages on sudden changes.

When Nord Pool has cleared the spot market and announced sales and purchases, all market players are required to submit schedules to the TSOs reflecting forecast demand and generation and all trades for each of the 24 hours of the following day. Trades can be both Nord Pool spot trades and bi-lateral trades. As Nord Pool has a monopoly on interconnection capacity, bi-lateral trades can only take place within a single zone. These schedules are binding in the sense that market players are financially responsible for their fulfilment. All market players with a physical footprint in terms of generation, consumption or trade after the scheduling deadline are required to register as balance-responsible market players. That is, they must sign a contract with the TSO in the zone in which they want this physical footprint; through this contract they become financially responsible for deviations and are bound to follow the specified rules and formats for communication with that TSO. Balance responsibility is the connection between the electricity market and the physical system responsibility held by the TSO. Various initiatives and efforts have tried to harmonise and even unify balance responsibility contracts within the Nordic TSOs, but some differences still remain.

After submitting schedules to TSOs, the framework deviates somewhat from country to country. In Sweden, Finland and eastern Denmark, it is possible to trade intra-daily up to a few hours before real-time operation. Nord Pool operates a trading platform, Elbas, for open intra-day auctions. So far, liquidity has been very low in this segment of the market. Balance-responsible market players in Sweden and Finland are also allowed to change their schedules until the final gate closure, when Elbas closes. There are also differences in real-time balancing. Nordic TSOs operate regulating markets in which they buy and sell electricity to balance the system according to the merit order of bids submitted by market players to TSOs. Prices for real-time regulation are determined by the marginal bid, as is the case in the day-ahead spot market. The Nordic regulating markets have become more and more integrated with enhanced information management systems. Now they function as a common Nordic market in which best bids and offers are called on across the entire Nordic region, subject to available transmission capacity.

The purchase of regulating resources is more or less harmonised, but the settlement of imbalances deviates on some critical points. The total imbalance of a system is the aggregate imbalances of all balance-responsible market players. A majority of the individual imbalances will be in the same direction as the aggregate system imbalance, but some imbalances will be in the opposite direction and will, thereby, actually decrease the aggregate system imbalance. These individual imbalances that, by chance, are actually helping the system are treated differently in the different Nordic countries. In all the Nordic countries, the balance-responsible market players that contributed to the imbalance are charged the same price as the marginal price from the purchase of regulating resources. In Norway, the imbalances that help the system by chance are also rewarded with the same price and are, thereby, treated equally with the market players that were actively called on in the purchase of regulating resources. This pricing principle is referred to as the "single-price" system and is cost neutral. Market players that caused the imbalance pay to those that alleviated the imbalance. The TSO is merely a co-ordinator or broker of the transactions.

In Sweden, Finland and Denmark, the balance-responsible market players that helped the system by chance are not rewarded. Their imbalances are settled with the day-ahead spot price, which always gives an equal or poorer remuneration than the price settled in the purchase of regulating resources. Otherwise, there would be an incentive to make arbitrage between the two markets. Thus, balance-responsible market players that caused the imbalance pay the settled regulating price to the TSO and the TSO passes this price on to those that were actively called on. In reality, they are settling imbalances of those that helped the system by chance at a less favourable price. This pricing principle is referred to as the "dual-pricing" system and is not cost neutral. In fact, it generates a profit for the system operator. It is intended to provide extra incentive for making good forecasts and maintaining balance, but in reality it creates an incentive to merge and decreases liquidity in the regulating market. If market players can cancel individual imbalances through merger and aggregation, they save the costs that would otherwise end as revenue to the TSO.

Imbalances are settled at the cleared regulating prices, usually one or two weeks after the day of operation. Local network companies collect hourly interval meter readings on a daily basis. These are matched with schedules to calculate individual imbalances. All Nordic countries have implemented systems for load profiling for the smallest consumers, primarily to avoid the need to have them install remotely read interval meters.

The system with balance responsibility that allocates financial responsibility amongst market participants, the system of firm transmission rights made available to the market and the *ex ante* price settlement could all be thought of as systems based on forced day-ahead contracts. Binding contracts are made prior to real-time operation and balance-responsible market players are forced into the intra-day and/or real-time adjustment market. In an integrated market, it is a system of forced long-term contracts (longer term than real-time) that enables optimisation and planning across several jurisdictions.

The day-ahead spot market is the cornerstone; real-time regulating and balancing markets provide the adjustment market between day-ahead spot and real-time operation. These markets create the interface between decisions made by market players and physical system operation, and could be called the markets with a physical footprint. However, most electricity trade and turnover takes place prior to the day of operation and is purely financial. Nord Pool spot has a market share of 43% of the physical Nordic demand; the remaining 57%

is traded bi-laterally. This could be thought of as bi-lateral physical trade but, in reality, it mainly reflects that several generators also have retail arms and therefore demand and generation are matched directly within the company.

Nord Pool also operates a trading platform for financial trading and a clearing house for bi-lateral, brokered, over-the-counter (OTC) trade. In the short term, Nord Pool offers contracts for one to nine days ahead and for one to six weeks ahead in time. These futures contracts are settled daily. In the long term, they have contracts for one to six months ahead in time, one to eight quarters ahead, and one to three years ahead. These longer forward contracts are settled at the end of the period during which they go to delivery. All these futures and forward contract use the daily average system price as reference. There are also contracts to hedge zonal price differences, either one quarter or one year ahead for which several competing electricity brokers hold the largest market share. Brokered contracts in the OTC market are still based on the same standards as the contracts marketed by Nord Pool. European Option contracts, in which the option can be exercised at a pre-specified date, are also traded at Nord Pool and OTC. The main part of the liquidity in this standardised option market is traded OTC. Nord Pool offers clearing of OTC-traded standard contracts. Volume and price data from OTC trading are published on Nord Pool's Web site, enhancing the flow of information in connection with its clearing service. This makes the Nordic market highly transparent, including the OTC market.

Market Structure

In 2004, total consumption in the Nordic market was 391 TWh with 146 TWh in Sweden, 122 TWh in Norway, 87 TWh in Finland and 36 TWh in Denmark. Some 379 TWh was supplied from Nordic production with 148 TWh coming from Sweden, 111 TWh from Norway, 82 TWh from Finland, 38 TWh from Denmark and the remaining 12 TWh from neighbouring markets. The bulk of the net-imports came from Russia. Net-trade with Germany can deviate substantially from year to year, depending on market fundamentals. In 2004, there was a net-export of some 2 TWh out of the total 11 TWh traded with Germany. In the same year, some 2 TWh were net-imported from Poland.

As of 31 December 2004, total installed capacity in the Nordic market was 91 076 MW. Some 47 059 MW was hydro capacity: 28 GW in Norway, 16 GW in Sweden and 3 GW in Finland. Hydro capacity in Norway is mainly in hydro reservoirs; in Sweden and Finland there is more run-of-river with less storing capability. The total maximum Nordic hydro storage capacity is some 120 TWh. Approximately 23 235 MW fossil-fuelled thermal capacity (mainly coal) was installed: 8 GW in Denmark, 6.5 GW in Finland and 4 GW in Sweden (most thermal capacity in Denmark was combined heat and power). About 12 142 MW of nuclear power was installed: 9.5 GW in Sweden and 2.5 GW in Finland. The last approximately 8.2 GW of power came from other renewable sources including 4 GW of bio-mass in Finland and Sweden, and 3 GW wind power in Denmark. Another 1 058 MW was added to the Nordic system in 2004 and 574 MW was taken off-line. The installed capacity served a Nordic peak demand of 66 GW.

Transmission interconnections between Nordic countries and regions, as well as between the Nordic market and neighbouring countries and regions, are a crucial part of the framework for a competitive Nordic market. Sweden and Norway are interconnected with 3 620 MW over nine different AC lines. Sweden and Finland are interconnected with 2 230 MW over five AC lines. Sweden and Denmark East are interconnected with 1 810 MW over four different sub-sea AC cables. Sweden and Denmark West are interconnected with 670 MW over two sub-sea DC cables. Norway and Denmark West are interconnected with a 1 000 MW sub-sea DC connection. Norway and Finland are interconnected with a single 100 MW AC line. The Nordic market is connected with neighbouring markets through 1 350 MW between Denmark West and Germany, 600 MW between Denmark East and Germany, 600 MW between Sweden and Germany, 600 MW between Sweden and Poland, and 1 560 MW between Finland and Russia. Several of the rated transmission capacities are subject to internal congestions and security requirements. As a consequence, some flows are reduced on a more or less permanent basis.

The four largest generating companies in the Nordic market are Vattenfall, Fortum, Statkraft and E.ON Sweden (formerly Sydkraft). In 2001 Vattenfall had a market share of 19% in terms of output compared to the total Nordic electricity generation. Vattenfall is owned by the Swedish state and has expanded with significant stakes into Finland, Denmark and Germany. Fortum had a market share of 16% in 2001. It is based in Finland with the Finnish state having a 60% stake; it recently made significant acquisitions in Sweden. Statkraft had a market share of 12% in 2001. It is owned by the Norwegian state. E.ON Sweden had a market share of 8% in 2001 and is owned by the German utility E.ON. With this, the four largest generation companies had a joint market share corresponding to 55% of the total Nordic generation. There has been a significant turnover of generation assets in Sweden, Finland and Denmark but very limited in Norway. Norwegian hydro resources are subject to restrictive concessions by the Norwegian state and this part of the market is not liquid.

No other companies held more than 4% of the market in 2001. In Norway, 160 companies are engaged in electricity generation; the 15 largest had 88%

of the market share of Norway's total generation in 2001. In Sweden, the 15 largest generators had a market share of 94% of the total domestic generation in 2001. In Finland, the 15 largest companies had a market share of 95%. The Danish market is undergoing rapid change. The two largest generators, Elsam and Energi E2, are merging but Vattenfall is taking over roughly half of the generation assets. Apart from this, there are hundreds of small combined heat and power plants and some 4 000 wind turbines. Distributed generation and wind power are being integrated into the market in the sense that they do receive a fixed subsidy per kWh produced output. Apart from that they have the responsibility to dispose of their output in the market - a responsibility previously placed on Danish TSOs.

In Norway, more than 100 retail companies are registered on the Competition Authorities' Web site, to serve some 2.5 million Norwegian consumers. Eighteen of these companies are registered as being retailers across the entire country. In Sweden, the Swedish Consumer Agency invites retailers to publish price information on their Web site. Some 80 companies have registered that they supply to the approximately 5 million final customers across the entire country; some 50 registered companies supply only to local customers. The Swedish Consumer Agency notes that almost all retailers that supply the entire country are registered, as are an increasing number of local suppliers. In Finland, there are roughly 80 retail suppliers for 3 million consumers. In Denmark, where there are 3 million consumers, some 80 to 90 retail suppliers are registered on a price information service; most of them are only local suppliers. Even though there appears to be a very large pool of retail companies in the Nordic market, a large majority of them are actually the retail arm of local, often municipally owned, distribution companies that supply only local customers.

At Nord Pool, some 400 participants are registered to trade in the day-ahead and/or financial market - either directly or through a handling agent. Roughly 150 of these are based in Norway, some 100 in Sweden and some 25 are based outside the Nordic region. The number of market participants has increased constantly since the launch of the market, up from some 150 participants in 1996.

Further Reading

Background and development of the Nordic electricity supply industry is described in IEA country reviews. The joint reviews of the Nordic countries since late 1980s constitute a comprehensive description of the development and of governing structures. These can be found on www.iea.org. Several descriptions of market design can be found on the Web sites of Nord Pool and the four Nordic TSOs; www.nordpool.com, www.statnett.no, www.svk.se, www.fingrid.fi, www.energinet.dk and www.nordel.org. These sites also contain comprehensive and detailed data and descriptions of the Nordic market's fundamental structure. A joint report by the Nordic competition authorities analyses the structure of this market in great detail (2003). Numerous research articles describe the development of the Nordic market and various market design aspects, including Flatabø *et al.* (2003) and von der Fehr *et al.* (2005).

ANNEX 3: AUSTRALIAN NATIONAL ELECTRICITY MARKET

Prior to the liberalisation process in Australia, the Australian electricity supply industry was dominated by state-owned utilities active in their own states. There was very little trade between states, except between Victoria and New South Wales (NSW) via the Snowy Mountain hydro system. Regulation of the electricity industry was under state jurisdiction. At the end of the 1980s, NSW, Victoria and Queensland had substantial overcapacity in generation, mainly due to optimistic projections of industrial demand growth made in the late 1970s and early 1980s. In the late 1980s, a range of initiatives were taken to address the various apparent inefficiencies in the Australian electricity industry. The Commonwealth Government Industry Assistance Commission was reviewing industry performance and pointed to a number of inefficient practices in an early stage. Some states were considering private construction and operation of generation plants and it was logical to consider building interstate transmission capacity to enable interstate trade. In March 1990, a new transmission line opened, connecting South Australia to the Victoria/NSW interconnected grid. Just two years later, as much as 24% of South Australia's electricity demand was met by power from the other states.

The liberalisation process in the Australian electricity market was effectively launched with the release of a draft report by the Commonwealth Industry Commission in January 1991. It recommended sweeping changes to the industry, with the main points being: placing transmission and distribution of gas and electricity on a commercial footing within one year; unbundling ownership of generation, transmission and distribution facilities within two years; breaking up electricity generating capacity into competing units; merging all state-owned transmission bodies into one organisation in order to establish a unified, national market; and requiring regulated third-party access to transmission and distribution networks. After privatisation, the final requirement was to bring tariffs in line with costs and remove cross-subsidies. A final report on the strategy was published in May 1991.

In July 1991, the heads of Commonwealth, state and territorial governments, through the Council of Australian Governments (COAG), agreed to establish an intergovernmental supervisory body, the National Grid Management Council, to encourage and co-ordinate efficient development of the electricity industry in eastern and southern Australia. The Council's mandate was to encourage open grid access, competition and free trade in electricity, and to co-ordinate planning of generation and interconnected transmission systems. The Council was formed as a co-operative body with an independent chairman; members were appointed by participating jurisdictions.

In the "One nation" statement in February 1992, Prime Minister John Major announced that he would seek states' agreement to establish a National Grid Corporation to operate the transmission network – separately from existing generation and distribution interests. At the same time, the Commonwealth offered to contribute up to AUD 100 million (app. USD 80 million) to upgrade the transmission network, subject to states' agreement to a timetable for developing the corporation. Additional financial incentives, totally AUD 4.2 billion up to 2005/06, were given by the federal governments. The Commonwealth followed up on the statement with a National Electricity Strategy discussion paper (May 1992) outlining the main elements of a policy to continue structural reform of the sector. In December 1992, the COAG agreed that the transmission sector should be managed separately from the generation sector. At the same time, the National Grid Management Council agreed to a protocol on the principles for development and operation of an interstate electricity grid.

Electricity market reform was only one of several sector reforms within a wider effort to improve the competitiveness of the Australian economy. In October 1992, the prime minister commissioned an independent inquiry on national competition policy following an agreement with COAG. This report, released in August 1993, proposed the abolition of a large number of regulations in all economic sectors and led to the first element of federal legislation (1995) that was to have a direct impact on regulation of the electricity industry. A key element in the legislation was the establishment of the Australian Competition and Consumer Commission (ACCC), which was to be involved in the regulation of the electricity sector.

In May 1994, COAG agreed to introduce a competitive national electricity market; by August 1994, they had agreed to the key principles and objectives of the market. These included unbundling, regulated third-party access, customer choice of supplier, regulatory reform of networks, merit order dispatch, non-discriminatory access for new entrants and no discriminatory barriers to inter- and intra-state trade. A key player in the National Electricity Market was the National Electricity Market Management Company (NEMMCO), the independent system and market operator. NEMMCO was founded in May 1996 with a membership comprising the governments of Queensland, New South Wales, Australian Capital Territory, Victoria, and South Australia.

Further details had been worked out by October 1996, when the National Electricity Code (NEC) was submitted to ACCC for approval. The NEC specifies market rules, network pricing, network connection and access, and system security. The National Electricity Code Administrator (NECA) was set up in May 1996 to enforce the code through implementation and monitoring of compliance.

At the state level, the necessary restructuring took place in parallel with the development of the key principles of the market, albeit in varying speeds. In Victoria, the electricity sector was privatised beginning in 1994. Generation assets were privatised, plant by plant, into six generating companies; distribution was privatised as five independent companies. Initially, the transmission company, PowerNet Victoria, remained public but was also privatised later. The first Australian power market started to operate in Victoria in 1994, managed by the Victorian Power Exchange. In NSW, transmission was unbundled and a market launched in 1996. The same happened in Queensland in 1998. In South Australia, the state-owned vertically integrated utility was corporatised in 1995 and unbundled by accounting in 1997. In 2000, its assets were put under private management through a long-term lease.

The first truly decisive and committing steps were taken in May 1996, when state energy ministers from NSW, Victoria, Queensland, South Australia and the Australian Capital Territory (ACT) signed an inter-governmental legislation agreement to support operation of the national electricity market in all participating jurisdictions. Ministers from NSW, Victoria and the ACT agreed to harmonise their existing electricity markets from 1 October 1996 to allow interstate competition at the earliest possible date, with expected market launch by the beginning of 1998. Queensland, South Australia and Tasmania were to join the market upon completing the necessary structural changes and connecting to the interstate grid.

The National Electricity Market commenced operation on 13 December 1998, including South Australia, Victoria, the ACT, NSW and Queensland.

After 2.5 years of operation (June 2001), it was decided to review market performance. This timing also coincided with a major reform of the England & Wales market, based on the obligatory pool system – which was somewhat similar to the key design principles of the Australian National Electricity Market. COAG agreed to establish the Ministerial Council of Energy (MCE), comprising energy ministers from the Commonwealth and all states and territories. MCE was to oversee the review process, also called the Parer review. The final report was published on 20 December 2002 under the name *Towards a truly national and efficient energy market*. Compared to development in England & Wales, where an overarching fundamental reform had resulted, the Parer review confirmed the robustness of the NEM's basic principles but pointed to various problems and gave recommendations on a wide range of necessary changes. One of the most important was the formation of two new bodies; the Australian Energy Market Commission (AEMC), which was to take over responsibility for rule making and market oversight from NECA; and the Australian Energy Regulator (AER), which would take over regulation from 13 state regulators. AEMC and AER were to have been established by July 2004 but the process was delayed to 2005.

Consumer contestability on the retail level remains an issue under full state responsibility. NSW and Victoria gave full retail contestability to all consumers from January 2002; contestability was granted in South Australia from January 2003 and from the ACT in July 2003. Queensland gave retail contestability for consumers of more than 100 MWh/yr from July 2004 but decided not to lower this barrier to include other consumers. This decision is based on a cost-benefit analysis, performed in 2001, claiming that the benefits for consumers would not outweigh the costs of enabling them to switch. NEMMCO operates the system that manages all retail switching, including the management of load profiles accepted for smaller consumers.

The NEM is now being extended to Tasmania through a new sub-sea DC cable connecting Tasmania with the interconnected national grid. Tasmania became a member of NEMMCO in May 2005, but will only be fully integrated in the market with the commissioning of the new cable, scheduled for early 2006. Retail contestability will develop in steps until the barrier is lowered to 150 MWh/yr in July 2009. At this point full contestability will be reviewed by the state.

Framework for Governance

The electricity supply industry is under the legal jurisdiction of states and territories, with the exception of issues related to interstate trade, which are in the jurisdiction of the Commonwealth as specified in the Australian constitution. The Constitution also specifies that all interstate trade must be free and fair. All necessary legislation that sets the framework for competitiveness and governance of the NEM in relation to interstate trade must be established in the legislations of states, territories and the Commonwealth. Close co-operation and co-ordination are an indispensable necessity for developing sufficient harmonisation to create a robust framework for governance.

State and territory legislation required unbundling of transmission and distribution networks from generation and retail. It has assigned rule-making and regulatory responsibilities to specific bodies and required third-party access. Transmission networks are unbundled by ownership. Detailed rule-making has been left to those joint bodies formed out of co-operation within the COAG. In practice, the National Electricity Law has been passed by parliaments agreeing on one state to take the lead, after which mirror legislations were quickly passed in the other states and territories and in the Commonwealth.

The new National Electricity Law appoints a new national body, the Australian Energy Market Commission (AEMC), to undertake rule-making and market development. The AEMC is responsible for the rules that determine market design, system operation and regulation through the new National Electricity Rules, which replaced the previous National Electricity Code. The AEMC reports directly to the Ministerial Council of Energy (MCE) and is governed by three commissioners. Its role is to make decisions on new proposed rules and to review the market, both on its own initiative and on that of the MCE. AEMC commenced operations from 1 July 2005, replacing the NECA. The National Electricity Law specifically requires the AEMC to establish a reliability panel that is to monitor, review and report on the security and reliability of the operation of the national electricity system.

In parallel with the establishment of the AEMC, the new National Electricity Law also appoints a new national body, the Australian Energy Regulator (AER), to perform economic regulation of the wholesale markets and transmission networks in the electricity and gas markets. The AER also ensures the compliance of the National Electricity Rules. The intention is also to transfer regulation of distribution and retail to the AER. The AER is an independent legal entity but a constituent part of the ACCC. It is governed by three commissioners, including a chairman, of which one must be a commissioner in the ACCC; the other two commissioners are appointed by states and territories (1) and by the Commonwealth (1). The rulings and decisions of the AER are subject to judicial review by the Federal Court of Australia.

In accordance with the National Electricity Rules, market and system operation are undertaken by the National Electricity Management Company Limited (NEMMCO), a company owned by states and territories that comprise the NEM. NEMMCO is governed by a board of directors, with each member appointing one individual. NEMMCO operates on a break-even basis by recovering the costs of operating the NEM from fees paid by market participants.

Basic Market Design Features

NEMMCO operates a spot market for wholesale electricity trade. The spot market is compulsory for all generation and consumption connected to the interconnected transmission grid that forms the back bone of the NEM.

Generators must submit offers and retailers can submit bids to NEMMCO for generation and consumption. These offers and bids must be submitted before 12.30 p.m., one day ahead of operation, specifying volumes and prices for each of the 48 half-hour periods in the following day. The volume in the bids can be changed up until five minutes before real-time operation, but the price cannot change. A pre-dispatch forecast of prices, volumes and dispatch is calculated for the following trading day. However, final prices and volumes are set only when the actual dispatch to meet demand at least cost has been determined - *i.e.* after the actual operation.

This *ex-post* market settlement has a wide range of implications for market operation, which are quite different from many other market systems. One implication is that nobody is forced to make binding commitments about generation, demand or transmission capacity prior to real-time. Without these forced commitments, there is no need for retailers to submit bids. As a default, NEMMCO makes load forecasts, which are used as input to make pre-dispatch forecasts. Retail companies can submit bids based on their own forecasts and particularly their own knowledge about demand response to prices. However, this approach makes little difference and has never really been used. In a system of *ex post* price settlement, it is not necessary to make the market two-sided. It does not, on the other hand, offer any implicit possibility to hedge activity prior to operation. All hedging must be performed through voluntary contracts between different market players. This has the advantage of minimising regulatory involvement and clarifying the interface between the responsibilities of NEMMCO and market participants. But it also has drawbacks in that transmission rights are not made firm, which makes it difficult to hedge for price differences and it makes the true independence and impartiality of transmission owners critical. It also creates challenges for demand participation: much demand will be far more responsive to price, if it is possible to plan the response one day ahead of operation. The NEM does not offer firmness for such planning. This should be developed in the normal market for hedging through aggregation and financial contracts.

Bids submitted by generators are stacked in ascending order. The generators with the lowest bids are dispatched until demand is met for every five-minute period. For each of these five-minute intervals, a marginal price is determined by the most expensive generator dispatched. A spot price is calculated as the average of the six marginal prices in the five-minute intervals over each half hour of the day. This average marginal price is then used to settle the market. Consumers and retail companies are charged this price for all consumption through this half hour; generators are paid this price for all dispatched generation through this half hour. Unit commitments and other factors related to the fixed costs of plants are not incorporated into the market price calculation. Bids of a certain volume and at a certain price are dispatched to the extent that they are necessary to meet load. On the other hand, as market participants are allowed to adjust the volumes of the bids until five minutes before operation, market participants are left some room to manage such constraints.

Transmission capacity is made available to NEMMCO by transmission system owners. Prices are calculated on a zonal basis, in which each member state is one zone and the Snowy Mountain region on the border between Victoria and NSW comprises an additional individual zone (due to the special role of its hydro capacity). Grid losses are assessed through computed loss factors at different locations in the grid, relative to specific reference points. Loss factors are calculated annually, based on historical data. Generators must compensate for grid losses and this is taken into account in the calculation of the dispatch schedule. In that sense, generators face a locational signal on two levels: one on the level of the zone, when taking congestion of interconnectors into account; and one on the level of distance to a zonal reference point, when taking grid losses into account.

NEMMCO must operate the system to balance supply and demand while at the same time maintaining a minimum reserve margin. The minimum reserve margin is currently set on levels equivalent to the capacity of the largest generators in each zone, but at the same time accounting for the possibility to share reserves across zones. The minimum reserve margin is required to enable the system to withstand the sudden failure of any generator. NEMMCO does not acquire operational reserves on a regular basis; however, NEM rules give NEMMCO the possibility to tender for reserves if they fall below minimum levels. In the beginning of 2004, the *Statement of Opportunities* predicted a reserve deficit of 356 MW below minimum during the summer months of February and March 2005, in South Australia and Victoria. NEMMCO initiated a tendering process to acquire operational reserves, but it was never necessary to enter into contracts for these reserves.

NEMMCO purchased ancillary services in eight parallel markets for frequency control, network control and black start. Competing service providers can bid their services into these eight markets. Financing of ancillary services is based on "causer pays" principles.

NEM rules also regulate a market for financial transmission rights. Price differences between zones give NEMMCO a congestion rent, also called settlement residues. Price differences create risks for market players and NEMMCO operates a financial market that facilitates hedging these risks. Rights to settlement residues are auctioned on a quarterly basis. Buying a certain share gives the right to the corresponding share of the settlement residues, but the right is not firm. If interconnection capacity is reduced or removed, the rights are lost with no compensation. Settlement residues and auction revenues are used to lower transmission use-of-service charges.

Transparency and information management are critical points in the NEM market design. NEMMCO releases an annual *Statement of Opportunities*, presenting and analysing all fundamental data on demand, transmission and generation. NEMMCO operates a Web site with very extensive dissemination of information. Prices, volumes, changes to the system and other fundamental market data are easily available. A unique feature is that all bids and offers submitted to NEMMCO are publicly available one day after operation.

Most trade in the mandatory spot market is backed by financial contracts (between market participants) to manage price and volume risks. Trade with financial contracts, apart from the settlement residue auction, is not regulated in the NEM rules. Financial trade takes place mainly in the bi-lateral OTC market, facilitated by brokers. Electricity derivatives are also traded on the Sydney Futures Exchange.

One segment of the financial market is instituted and regulated by the NSW government as a forced hedging contract. On 1 January 2001, the NSW government instituted the Electricity Tariff Equalisation Fund (ETEF), which requires electricity retailers in the state to pay money into a fund when the NSW pool price is below a certain, regulated threshold. If the pool price is above the threshold, retailers receive money instead. In that sense, the ETEF works as a protection that allows retailers to earn a guaranteed margin. But it also undermines competition and incentives for innovation. If the fund develops a negative balance, state-owned generators must contribute funds to keep the fund solvent. Generators are repaid when pool price developments recover the balance through payments from retailers.

Market Structure

In 2003, electricity supplied to the 8 million consumers in the NEM was 158 479 GWh, of which 65 TWh (41%) was generated in NSW and the ACT,

40 TWh (25%) in Victoria, 42 TWh (27%) in Queensland and 12 TWh (7%) in South Australia. The introduction of Tasmania into the NEM will increase load with some 10 TWh. Peak demand in the summer 2003/04 was 12.2 GW in NSW and the ACT, 8.6 GW in Victoria, 7.9 GW in Queensland and 2.6 GW in South Australia. Total NEM peak load was 29.8 GW, including the 1.3 GW in Tasmania, which is 2.9 GW less than the sum of the regional peaks.

In 2003, demand in the NEM was met by generation from coal (91%), natural gas (6%) and hydro (3%). In the same year, NSW imported 8 TWh, Victoria was a net exporter of 4 TWh, Queensland exported 3 TWh, South Australia imported 2 TWh and the Snowy Mountain Region exported 4 TWh of the 5 TWh generated in that region. By the end of 2003, total installed principal generation capacity in the NEM was 38 533 MW of which 12 GW (32%) was in NSW, 8 GW (22%) in Victoria, 11 GW (28%) in Queensland, 3 GW (9%) in South Australia and 4 GW (10%) in the Snowy Mountain Region. Extending the NEM to Tasmania will increase installed capacity with 2.5 GW of mainly hydro capacity.

Interconnections between various regions are the backbone of the NEM's integrated national market. NSW and Queensland are interconnected with 950 MW transmission capacity, of which only some 600 MW are made available for imports to Queensland. NSW and the Snowy Mountains are interconnected with 3 038 MW. Victoria and the Snowy Mountains are interconnected with 1 892 MW. Victoria and South Australia with 627 MW, of which only 379 MW are made available for imports to Victoria. Victoria and NSW are interconnected via the Snowy Mountains, but imports to the Snowy Mountains are limited to some 1 100 MW from either side.

Generation assets in the NEM are owned both by private and state-owned companies. The three largest generation companies in terms of installed capacity are Macquarie Generation, Delta Electricity and Eraring Energy. All are based in NSW and owned by the NSW government. They dominate the NSW region with shares between 40% and 25% of the total installed capacity. In the NEM, they have between 12% and 8% of the market share and a joint market share of 31%. In total, there are 27 generation companies in the NEM, one-third of these are state-owned. Apart from NSW, Queensland also owns the dominating generation companies in that state. All generation assets in Victoria and South Australia are operated by privately owned companies, several of which are non-Australian.

On the retail side, there are 66 retailing companies supplying the 3.2 million consumers in NSW and the ACT, the 2.3 million consumers in Victoria, the 1.7 million consumers in Queensland and the 0.8 million consumers in South Australia.

By mid-2005, NEMMCO had registered 103 market participants; six of these were engaged only in trading.

Further Reading

Background and development of the electricity supply industry in Australia is described in IEA country reviews. The joint reviews of Australia since the late 1980s constitute a comprehensive description of the development and of governing structures. The latest country review from 2005 gives a comprehensive description of the structure of the Australian electricity industry. These can be found on www.iea.org. The Web site of NEMMCO, the system and market operator, also contains a wealth of information, data and descriptions on most aspects of the NEM: www.nemmco.com.au.

ANNEX 4: PENNSYLVANIA – NEW JERSEY – MARYLAND INTERCONNECTION (PJM)

The Pennsylvania – New Jersey – Maryland Interconnection (PJM) has a much longer history than that of the development of electricity market liberalisation itself. PJM has been a pool that enables co-ordination of trade between the three founding utilities since 1927.

Prior to 1978, the United States electricity industry was run by vertically integrated utilities, in most cases privately owned. These companies were regulated by the state public utilities commissions (PUCs). The Federal Power Act of 1935 kept most regulatory authority in the hands of the states. The Public Utility Holding Company Act of 1935 (PUHCA) limited the possibility for any utility to do business in other states, thereby also effectively limiting the scope for mergers and acquisitions. On the federal level, the Federal Energy Regulatory Commission (FERC) has authority only over wholesale trade issues. An important development away from this approach was triggered by the Public Utility Regulatory Policies Act of 1978 (PURPA), which gave small generation resources – particularly with environmental merits – access to the grid through contracts corresponding to avoided costs. This led to an influx of independent power producers (IPPs), primarily in those states where the vertically integrated utilities were permitted and encouraged to auction least-cost contracts to IPPs to obtain the power needed.

The next important step towards a more competitive market was taken on a federal level with the Energy Policy Act of 1992. This act gave the FERC authority to order open access for wholesale trade between utilities and across state borders. It specifically prohibited the FERC from ordering open access for final consumers. This was still a matter for state legislation.

In anticipation of developments the Energy Policy Act of 1992 would trigger, PJM started to transform itself into an independent, neutral organisation in 1993, primarily through the formation of PJM Interconnection Association, which administered the power pool. By 1995, the FERC had developed a draft set of rules that would implement the Energy Policy Act of 1992. In 1996, the final set of rules was issued in Order 888. The key points in the FERC Order 888 include: functional unbundling of utilities' transmission system operation business from their power marketing business; transmission utilities under FERC jurisdiction must file non-discriminatory, open-access transmission tariffs that offer service to third parties on a comparable basis to the utilities' own uses of their transmission facilities; and guidelines for independent system operators (ISOs) that would be subject to FERC jurisdiction. In 1999, the FERC issued order 2000, which asks for consolidation across jurisdictional borders, meaning that utilities are encouraged to merge ISOs into regional transmission organisations (RTO) that act as co-ordinating system operators across larger areas.

In September 2001, the chairman of the FERC made several proposals to encourage the standardisation of market design and push for the formation of RTOs. The proposals were clarified in a standard market design (July 2002) and the FERC issued a white paper (April 2003) with a refined version of the proposals. After strong resistance from some states, the proposals have never materialised as an instrument to force wholesale competition in all states. In the Energy Act of 2005 (July 2005), the FERC was given some of the authority initially suggested in its standard market design. With this act, the FERC has much stronger authority in matters of system security and it has been given some authority in the approval process of new transmission infrastructure and to monitor and enforce competitive behaviour in wholesale markets. In short, the Energy Act of 2005 indicates that the development towards competitive and open electricity markets should be supported, but not enforced on all states. The Energy Act of 2005 also abolishes the limitations set by the PUHCA, thereby opening up for consolidation across states.

With the legal framework of Order 888 and the intentions expressed by the FERC in mind, PJM became a fully independent organisation in 1997. Ownership was opened to non-incumbent utilities and an independent Board of Managers was elected. The original PJM Interconnection comprised the control areas of eight utilities, covering large proportions of Pennsylvania, most of New Jersey, large portions of Maryland and Delaware in its entirety. A bid-based spot market for power commenced operation on 1 April 1997, one year before the launch of the Californian market. Later in 1997, PJM was approved by the FERC as the first ISO in the country to be in compliance with Order 888. PJM was thereby responsible for safe and reliable operation of the unified transmission system and for the management of a competitive wholesale electricity market across the control areas of its members.

With the FERC's encouragement of the formation of RTOs in mind, PJM, the New York ISO (NYISO) and the ISO New England (ISO-NE) tried to agree to merge to form a unified RTO. However, an agreement could not be struck and in December 2002, PJM earned full RTP status on its own. The period between the issuance of Order 2000 and the approval of PJM as an RTO also covers a period with substantial turbulence as more and more states embarked on the electricity market liberalisation process. California was one of the first states to order full retail access in March 1998: a competitive wholesale market was launched on 1 April 1998 and performed very well initially. It reinforced the merits of liberalisation but the subsequent poor performance from 2000 and the eventual crash in summer 2001 had an even stronger negative influence on the liberalisation process in the US.

The first years of PJM operation were used to firmly establish and develop the market to become a robust framework for competition. The initial day-ahead spot market was based on a single market-clearing price for the entire region. High costs for congestion management and poor operational flexibility in the utilisation of the system (due to security restrictions) called for a stronger locational reflection of real costs. One year after the initial launch, the level of sophistication increased by introducing locational marginal pricing (LMP) based on reported costs, in which market-clearing prices were calculated for each node in the system. On 1 January 1999, a daily capacity market was introduced, motivated by the traditional approach to capacity planning in the region and by a concern for stranded costs from incumbents, who would now be faced with competition from generators without former times' capacity requirements. From March 1999, the daily capacity auctions were extended to monthly and multi-monthly auctions. On 1 April 1999, the cost-based LMP was replaced with LMP based on real competitive bidding. On 1 June 2000, the day-ahead market was extended with a real-time market, also based on LMP and competitive bidding. On December 1 2000, a market for spinning reserves was added. With the implementation of LMP principles in 1999, there appeared a need to offer hedging of the consequential price differences between nodes. In April 1999, PJM introduced an auction of allocated financial transmission rights (FTRs), which gave market participants the needed risk-hedging opportunity. In May 2003, the sophistication of this market increased again by replacing the initial allocation of FTRs with an entirely financial allocation of auction revenue rights (ARRs).

By the end of 2001, the market design had developed to create a framework for a robust market. The market design has continued to develop since then, particularly through modifications of the capacity market. Since 2002, there has been a focus on extending PJM system and market operation area. On 1 April 2002, Allegheny Power joined PJM to form PJM West. The extension added more regions of Pennsylvania, large parts of West Virginia, parts of Virginia and small parts of Ohio. In June 2002, American Electric Power (AEP), Commonwealth Edison (ComEd), Illinois Power and National Grid signed a memorandum of understanding with PJM to develop an independent transmission company that would operate within PJM West. Also in June

2002, Dominion joined PJM to form PJM South, thereby integrating a large share of the electricity system in Virginia and a small share in North Carolina into PJM's system and market operation. AEP (Ohio), Com Ed (Illinois), the Dayton Power & Light Company (Ohio), Duquesne Light Company (Pennsylvania) and Dominion were successfully integrated into the system and market operation of PJM during 2004 and first half of 2005. The integration of Com Ed alone expanded the PJM market by 20%.

Midwest ISO (MISO) and PJM have worked together since 2004, with the goal of developing an integrated wholesale market across the 23 states, the District of Columbia and Manitoba (Canada), which covers their joint control areas. MISO launched a LMP-based market on 1 April 2005.

Most of the states in PJM have ordered retail access for all consumers. The first state in the United States to order full retail access was Rhode Island (January 1998), a part of ISO-NE. The first state in PJM to follow suit was New Jersey (August 1999). Pennsylvania, D.C., Delaware, Ohio, Maryland and Illinois gave full retail access at various points between 2000 and 2004. By 2004, 18 states had given retail access to all consumers.

Framework for Governance

Three major federal acts on energy determine the level of involvement of federal authority in the regulation of the electricity industry in the United States. The Federal Power Act of 1935 limits the federal authority to issues related to wholesale trade. The Energy Policy Act of 1992 ordered open access for wholesale transactions, meaning that there must be open access to trade between utilities within and across states. Finally, with the Energy Policy Act of 2005, the intentions laid out in the Energy Policy Act of 1992 were confirmed and federal authority was strengthened in some critical areas such as system security, transmission system extensions and market monitoring. Some state legislation, including in most of the states within PJM, gives full retail access to final consumers.

State regulators, together with a federal regulator, oversee compliance of state and federal legislation. States have public utility commissions (PUCs) and the Federal Energy Regulatory Commission (FERC), which is an independent agency within the Department of Energy, regulates on those areas in which federal legislation gives it authority. PUCs regulate intra-state utility business, such as generation and distribution. The FERC regulates interstate energy transactions, including wholesale power transactions on transmission lines. The FERC consists of five members appointed by the

President and confirmed by the Senate. No more than three commissioners may belong to the same political party. The President appoints its chairman.

In Order 888, the FERC established the set of detailed rules that must be fulfilled to comply with the Energy Policy Act of 1992.

In the 1960s and 1970s, large blackouts raised the issue of secure and reliable operation of interconnected transmission systems that reach across state and jurisdictional borders. In 1968, the North American Electric Reliability Council (NERC) was formed by the 10 regional reliability councils that had been formed in preceding years. The NERC is a voluntary organisation with the mission of ensuring that the North American electricity transmission grid is reliable, adequate and secure. The lessons from the widespread northeastern blackout in 2003 pointed to the inadequacy of voluntary reliability rules, particularly concerning lack of enforcement of otherwise appropriate rules. In the Energy Policy Act of 2005, the FERC was given authority to appoint a body to act as regulator of a set of mandatory reliability rules.

PJM Interconnection is a limited liability, non-profit company, governed by a board of managers. Members of the board of managers must have no personal affiliation or ongoing professional relationship with - or any financial stake in - any PJM market participant. Users of PJM join as members and are represented with a vote in the members committee. The members committee elects a board of managers and provides advice to this board by proposing and voting on changes in market rules; it also has authority to make specific recommendations. There are other committees and user groups for resolution of issues through discussion and negotiation. Market rules and market design issues are often developed through these governing structures.

There is a specific unit within PJM to oversee the functioning of the market; the Market Monitoring Unit (MMU). The MMU is an independent group that assesses the state of competition in each of PJM markets, identifies specific market issues and recommends potential enhancements to improve competitiveness and market efficiency. In particular, the MMU is responsible for monitoring the compliance of members with PJM market rules and for evaluating PJM policies to ensure those rules remain consistent with the operation of a competitive market. The MMU issues an annual report on the state of the market.

Basic Market Design Features

It costs resources to transport electricity, both in terms of keeping adequate transmission assets available and in terms of energy losses due to resistance in transmission grids. In PJM, this fact has been taken very explicitly into account by introducing a nodal pricing system, which means that a price is calculated in every possible choke point in the electricity system. Every bus that connects two transmission and/or distribution lines is a possible choke point and is, thus, incorporated directly into the calculation of prices.

All generators defined as a capacity resource in PJM system are obliged to submit an offer into the day-ahead PJM market. The bus that connects a generator to the grid is specified when registering. Offers can include incremental prices, specifying different prices at different generation volumes. They can also specify minimum run times and start-up costs to ensure that unit commitments are incorporated into the market clearing price. Market participants are allowed to self-schedule. On the generator side this is accounted for by indicating that a specific share of a generation unit must run regardless of the price. An offer specifying that a unit must run is basically just a schedule that commits the generator financially.

Retailers and consumers must submit bids to the day-ahead spot market. They can do it by bidding prices and volumes, if they intend to respond to the price by decreasing demand, or they can do it without specifying any price.

All demand bids and generation offers are tied to a specific bus and transmission capacity between all buses are specified. Load and/or generation at a specific bus will have a specific influence on energy losses from transmission. These losses will be dynamic but sensitivity factors are calculated annually as approximations of the transmission losses they incur relative to a reference point in the grid. With all bids, offers and sensitivity factors, a dispatch schedule is calculated that minimises total system costs of serving total demand. Reliability and transmission system security considerations are taken into account in the total market clearing. A marginal pricing principle is used, whereby the most expensive offer that meets the cheapest bid sets the price. This nodal pricing is the instrument of congestion management internally within PJM market area. Interconnections between the PJM area and other neighbouring areas are also defined as nodes. External bids and offers across these nodes determine the price in the node and the flow across the interconnector.

Other details are also included in the market clearing calculation. Certain generators are included in the price calculation, according to a previously agreed "cost of service" due to their position and power in the market. Other generators might be forced to operate due to reliability and stability criteria in the operation of the system. Depending on their location, these generators are not always included in the calculation of locational marginal nodal prices, according to their offers.

Market clearing prices are calculated at all buses; if there are congestions between nodes, nodal prices may be different. These price differences reflect the cost of electricity transmission. Each generator is paid the market clearing price in its specific node. All loads are charged the market clearing price in their specific nodes. For market participants that have chosen to self-schedule, either by letting own generation meet own load or by entering into other bi-lateral contracts, the resulting price differences are exactly the charges that reflect the costs connected to a specific location. In 2004, some 26% of the load was cleared in PJM day-ahead market. The remainder was generation offers submitted as must run, meaning it was self-scheduled.

Price differences between nodes can imply a significant market risk for market participants. At the same time, these price differences generate revenue streams to PJM known as financial transmission rights (FTRs). Owners of transmission grids and holders of contracts linking generation in one location with load at another can request the allocation of FTRs. Investors in new transmission interconnectors earn the right to FTRs tied to that line. FTRs can be traded, either bi-laterally or through PJM, which conducts annual and monthly auctions. A holder of an FTR for the flow of a specific volume in a specific direction between two nodes will receive the revenues of price differences between those two nodes - or will be charged if the revenue stream is negative. This creates a hedge for a contract that was made at a price referring to one of those nodes. Instead of tying the initial allocation of FTRs to specific congestion points, according to the earned rights, an initial annual allocation of auction revenue rights (ARRs) has been instituted. ARRs give the rights to the proceeds of an annual auction of FTRs. If the holder of the ARR has a specific interest in the underlying FTR, the ARR can be transformed into the corresponding FTR outside of the auction. With ARRs, the initial auction of FTRs has become more liquid and, hence, more flexible.

PJM spot market is a two-tier market with a day-ahead market and a real-time market. The pricing principles in the real-time market follow the same principles of nodal-based, locational marginal pricing as the day-ahead market. The real-time market is an adjustment market, in which deviations between scheduled bids and offers submitted the day-ahead are traded. Market participants are free to self-schedule to compensate for any deviations. PJM real-time market is a single-price market. Deviations that contribute to the total system imbalance are faced with the costs of regulation to achieve system balance. Deviations that, by chance, help the total system balance are faced with the same price as those resources that were actively called on to regulate. There is no extra punishment for imbalances in excess of the real costs of managing the imbalance.

PJM operates a market for installed capacity. All load-serving entities (LSEs: retail companies with contracts to serve consumers) are obliged to contract for an amount of installed capacity that corresponds to the peak load of the customer base plus reserves. The obligation must be met by capacity which is registered by PJM and available to the system. Owners of such capacity receive a capacity credit. LSEs can either contract for capacity credits bi-laterally or through the market for capacity credits organised by PJM. If LSEs are not meeting their obligation, they are charged with a fine, currently at a level of some USD 170 /MW and per day of unfulfilled requirement. PJM conducts daily, interval, monthly and multi-monthly auctions of capacity credits.

PJM also acquires ancillary services through market-based mechanisms. There is a market for capacity to meet regulation on a very short-term, automatic regulation. There is also a market for spinning reserve to make sure there is adequate capacity on-line to meet sudden failures in the system. Both markets are cleared every hour.

Markets operated by PJM include: the day-ahead spot market, which also constitutes a reference; the real-time spot market; a market for capacity credits; and two markets for ancillary services. A substantial part of electricity trade in the PJM market area takes place outside of the official PJM market places. Much load is met through self-scheduled generation, either within a vertically integrated company (generation and retail) or through bi-lateral contracts. Financial trading is evolving at the New York Mercantile Exchange (NYMEX) and the Intercontinental Exchange (ICE) where liquidity in terms of traded volumes is increasing rapidly, particularly over the past year (2004/05). There are more than 7 000 nodes for calculation of nodal marginal prices in the PJM model. Most of these nodes are never singled out with a specific price because they are not congested. PJM calculates weighted average prices for three hubs that consist of important non-congested nodes. These hubs are used as references for risk management and financial trading. The most important hub is the western hub, which attracts the most liquidity as a reference point.

PJM Web site offers extensive information on market prices, the state of the electricity system, principles, governance and other valuable information to understand the overall structure of the market and its day-to-day evolution. Data showing individual bids, without identifying the bidder, are published with a six-month delay. There is no requirement or corresponding site that contains information regarding the status of specific generation units. A generation unit can be taken off-line – with possible significant consequences for involved nodal prices – without immediate pass through of that information to the market place.

Market Structure

Statistical data on demand and supply in the PJM market has changed substantially during the last years with the rapid extension of the market area. In 2004, there where three important phases: Phase 1 with 12 control zones; Phase 2 where PJM also comprised the ComEd control zone, and; Phase 3 where PJM also included American Electric Power and Dayton Power & Light. In 2005, PJM was further extended to also include Duquesne Light Company and Dominion.

In 2004, demand peaked at 78 GW in the extended PJM market. Assessed peak demand after the extensions in 2005 is 130 GW; total demand is some 700 TWh. The joint population in the area, after the 2005 extensions, is 51 million spread across 13 states plus the District of Columbia.

During Phase 1, PJM was a net importer: during the first four months of 2004, some 7.2 TWh were imported to PJM. In the extension of PJM in Phases 2 and 3, PJM became a net exporter with a net export of some 5.5 TWh during five months in Phase 2 and some 3.9 TWh of net export in the last three months of the year. This made PJM a net exporter of some 2.2 TWh for the calendar year 2004. The net-export and total load was served by 144 GW of installed production capacity as of 31 December 2004: 41.5% was coal-fired; 28.4% was fuelled by natural gas; 19.1% was nuclear; 7% was oil; 3.7% was hydroelectric; and 0.3% was solid waste. In the calendar year 2004, the installed coal units generated 52.1% of total generation; 36.9% was from the nuclear stations; units fuelled by natural gas generated 7%; oil was 1.1%; hydroelectric generated 2.3%; and solid waste generated 0.6%. With the extensions in 2005, the total installed generation capacity increased to 164 GW.

With the inclusion of ComEd, a special feature was introduced in PJM market, which is particularly important when analysing the market structure. ComEd is connected to the other PJM area via a pathway through a geographic area that is not a part of PJM market.

By the end of 2003, American Electric Power Company was the largest generation company in PJM, owning 17% of the total installed capacity and generating 22% of the total output that year. Exelon was second with 13% of installed capacity and 23% of the total generation. Third was Public Service Enterprise Group (PSEG) with 9% of installed capacity and 6% of generated electricity. Exelon and PSEG have proposed to merge. In the market-screening process of the consequences of the merger, PJM Market Monitoring Unit has analysed the resulting market shares. After a merger, it is indicated that the market share in terms of installed capacity will result in a merged company with a 29% market share, up from 22%, as well as another company with 22% and at least three companies with market shares of 6% each. It is noted that a forced divestiture of 4 500 MW would reduce market concentration to pre-merger levels.

The market for retail supply is overseen by state public utility commissions. In states with customer choice, there are lists of licensed suppliers from which consumers can choose. In Pennsylvania, there are 40 licensed suppliers of which roughly half are registered as suppliers to all types of customers; the remaining serve only industrial and commercial customers. A similar picture can be found in other PJM states such as New Jersey and Maryland.

PJM has more than 375 members with generation capacity and/or retail supply commitments, or with only trading commitments.

Further Reading

The fact that electricity markets in the United States develop geographically on a control area basis, with the extension of markets proceeding control area by control area subject to state regulation, makes it complicated to get a full overview of the development. Hunt (2002) provides a thorough description and discussion of the development in the United States, including PJM. PJM Web site (www.pjm.com) provides substantial amounts of material describing the various aspects of the market model. The Market Monitoring Unit publishes an annual report on the state of the market, as does the FERC, presenting thorough analysis of all markets in the United States. Many research articles have been published on the Californian crisis, the various market design aspects of PJM (including LMP and capacity markets) and on market power analysis.

REFERENCES

Alberta Department of Energy, 2005, "Alberta's Electricity Policy Framework: Competitive - Reliable - Sustainable", Review of Alberta's electricity market, www.energy.gov.ab.ca

Australian Government, 2004, "Securing Australia's Energy Future", www.pmc.gov.au/initiatives/energy.cfm, Commonwealth of Australia, Canberra

Boot, Pieter, 2005, "Security of Electricity Supply - the Dutch Approach", presentation at IEA/NEA workshop *Security of Energy Supply for Power Generation - May 2005*, www.iea.org

Bushnell, James B., & Catherine Wolfram, 2005, "Ownership Change, Incentives and Plant Efficiency: The Divestiture of US Electric Generation Plants", CSEM Working Paper 140, University of California Energy Institute

Centre for Advancement of Energy Markets, 2003, "Estimating the Benefits From Restructuring Electricity Markets - An Application to PJM Region", www.caem.org

Council of Australian Governments Energy Market Review, 2002, "Towards a truly national and efficient energy market", Commonwealth of Australia

CPB, 2004, "Energy Policies and Risks on Energy Markets - A cost-benefit analysis", CPB Netherlands Bureau for Economic Policy Analysis, The Hague

Cramton, Peter & Steven Stoft, 2005, "A Capacity Market that Makes Sense", conference paper, EPRI conference on Global Electricity Industry Restructuring - in Search of Alternative Pathways, May 2005, http://www.lrca.com/events/EPRI-AC/

Danish Competition Authority, 2003, "Elsam og Energi E2 afgiver tilsagn så konkurrencen fremmes" ("Elsam and Energi E2 gives consent that supports the improvement of competition"), press release, www.ks.dk

ECON Energy, 2003a, "2002: Høsten da tilsiget sviktet", www.econenergy.com

ECON Energy, 2003b, "Torrår i Norden - Vad hände på kraftmarknaden", www.econenergy.com

ENEL, 2005, "ENEL Annual Report 2004", ENEL financial statement, www.enel.it

ESAA, 2002,"Electricity Australia 2002", Electricity Supply Association of Australia Limited, Melbourne

ESAA, 2004, "Electricity Australia 2002", Electricity Supply Association of Australia Limited, Melbourne

ETSO, 2004, "An Overview of Current Cross-border Congestion Management Methods in Europe", ETSO report on congestion management, www.etso-net.org

ETSO & EuroPEX, 2004, "Flow based market coupling", joint ETSO and EuroPEX proposal, www.etso-net.org

European Commission, 2000, "Commission allows merger of VEBA and VIAG subject to stringent conditions", Press release IP/00/613 on 13 June 2000, www.europe.eu.int, Brussels

European Commission, 2004a, "Trans-European Energy Networks: TEN-E Priority Projects", European Commission DG for Energy and Transport, www.europe.eu.int, Brussels

European Commission, 2004b, "EU Productivity and Competitiveness: An industry Perspective", European Commission DG Enterprise, www.europe.eu.int, Brussels

European Commission, 2005, "Report from the Commission on the Implementation of the Gas and Electricity Internal Market: Technical Annexes", *Commission Staff Working Document*, European Commission DG for Energy and Transport, Brussels.

EFET-European Federation of Energy Traders, 2004, "Reforming the Management of Electricity Transmission Congestion in the EU Internal Market: an EFET Vision", EFET position paper, www.efet.org

European Parliament, 2005b, European Parliament legislative resolution on the proposal for a Directive of the European Parliament and of the Council Concerning Measures to Safeguard Security of Electricity Supply and Infrastructure Investment", www.europe.eu.int, Brussels

European Union, 2003a, "Directive 2003/54/EC of the European Parliament and of the Council of 26 June 2003 Concerning Common Rules for the Internal Market in Electricity and repealing Directive 96/92/EC", *Official Journal of the European Union*, L 176/37-55, Brussels

European Union, 2003b, "Regulation (EC) No 1228/2003 of the European Parliament and of the Council of 26 June 2003 on conditions for access to the network for cross-border exchanges in electricity", *Official Journal of the European Union*, L 176/1, Brussels

Evans, Joanne & Richard Green, 2005, "Why did Brittish Electricity Prices fall After 1998", *Department of Economics Working Paper*, WP 05-13, University of Birmingham

Federal Energy Regulatory Commission (FERC), 1996, "Order No 888, Promoting Wholesale Competition through Open Access", www.ferc.gov

von der Fehr, Nils-Henrik, Eirik S. Amundsen and Lars Bergman, 2005, "The Nordic Market: Signs of Stress", *The Energy Journal*, special issue on European Electricity Liberalisation

FERC, 2005, "State of Markets Report 2004", Annual FERC energy market report, www.ferc.org

Flatabø, Nils, Gerard Doorman, Ove S. Grande, Hans Randen and Ivar Wangensteen, 2003, "Experience With the Nord Pool Design and Implementation", *IEEE Transactions on Power Systems*, vol. 18, no. 2

FSE-Foreningen for Slutbrugere af Energy, 2003, "Klage til EU Kommissionen over svensk stop for dansk import af elektrisk kraft fra Sverige" ("Complaint to the EU Commission over Swedish blocking of Danish import of electric power from Sweden"), letter from FSE to the European Commission, www.fse.dk

Green, Richard, 2004, "Electricity Transmission Pricing: How Much Does it Cost to Get it Wrong?", *Cambridge Working Papers in Economics*, CWPE 0466, The Cambridge-MIT Institute

Holmen, 2005, "Holmen signs electricity supply and energy efficiency agreements with Vattenfall", press release, www.holmen.com

Hunt, 2002a, "Making Competition Work in Electricity", John Wiley & Sons, New York

IEA, 2002a, "Distributed Generation in Liberalised Electricity Markets", IEA/OECD, Paris

IEA, 2002b, "Security of Supply in Electricity Markets", IEA/OECD, Paris

IEA, 2003a, "Power Generation Investment in Electricity Markets", IEA/OECD, Paris

IEA, 2003b, "The Power to Choose - Demand Response in Liberalised Electricity Markets", IEA/OECD, Paris

IEA, 2004a, "Security of Gas Supply in Open Markets - LNG and Power at a Turning Point", IEA/OECD, Paris

IEA, 2004b, "World Energy Outlook 2004", IEA/OECD, Paris

IEA/NEA, 2005, "Projected costs of generating electricity - 2005 Update", OECD/IEA Paris

IEA, 2005a, "Russian Electricity Reform: Emerging Challenges and Opportunities", IEA/OECD, Paris

IEA, 2005b, "Energy Policies of Spain", IEA/OECD, Paris

IEA, 2005c, "Learning from the Blackouts: Transmission system security in competitive electricity markets", IEA/OECD Paris

IEA, 2005d, "Act Locally, Trade Globally: Emissions trading for Climate Policy", IEA/OECD, Paris

Joskow, Paul L., 2003, "The Difficult Transition to Competitive Electricity Markets in the US", *Electricity Restructuring: Choices & Challenges*, eds. J. Griffin and S. Puller, University of Chicago Press.

Joskow, Paul L., 2004, "Integrating Transmission Networks and Competitive Electricity Markets", presentation at IEA workshop *Transmission Network Performance in Competitive Electricity Markets*, www.iea.org

Joskow, P. & Jean Tirole, 2004, "Reliability and Competitive Electricity Markets", Massachusetts Institute of Technology Department of Economics Working Paper Series, Working paper 04-17

Littlechild, S., 2004, "Regulated and Merchant Interconnectors in Australia: SNI and Murraylink Revisited", *Cambridge Working Papers in Economics*, CWPE 0410, The Cambridge-MIT Institute

Ministry of Petroleum and Energy, 2003, "Om forsyningssikkerheten for strøm mv.", Stortingsmelding nr. 18, www.oed.no

NEMMCO, 2004, "Statement of Opportunities for the National Electricity Market 2004"

NEMMCO, 2005, "Statement of Opportunities for the National Electricity Market - Update 2004"

Newbery, D. & M. Pollit, 1997, "The Restructuring and Privatisation of Britain's CEGB - Was it worth it?", *The Journal of Industrial Economics*, vol. 45, No 3, pp. 269-303, Blackwell Publishing Ltd

Newbery, D., R. Green, K. Neuhoff, P. Twomey, 2004, "A Review of the Monitoring of Market Power", Report prepared at the request of ETSO, www.etso-net.org

Newbery, David, 2005, "Electricity Liberalisation in Britain", *The Energy Journal*, special issue on European Electricity Liberalisation

Nordel, 2003a, "Action plan: Peak Production Capability and Peak Load in the Nordic Electricity Market", Nordic Council of Ministers and Nordel

Nordel, 2003b, "Power and energy balances, today and three years ahead", www.nordel.org

Nordel, 2004a, "Peak Production Capability and Peak Load in the Nordic Electricity Market", Summary report in the Nordel project on Peak Load Capability and Demand in the Nordic Electricity Market, www.nordel.org

Nordel, 2004b, "Demand Response in the Nordic Countries - A background survey", A working document in the Nordel project on Peak Load Capability and Demand in the Nordic Electricity Market, www.nordel.org

Nordel, 2004c, "Annual report 2003", www.nordel.org

Nordel, 2005a, "Enhancing Efficient Funtioning of the Nordic Electricity Market", www.nordel.org

Nordel, 2005b, "Nordel to Finance Jointly Network Investments and to Streamline Market Rules", Nordel press release 28 February 2005, www.nordel.org

Nordic Competition Authorities, 2003, "A Powerful Competition Policy: Towards a More Coherent Competition Policy in the Nordic Market for Electric Power", www.ks.dk, Copenhagen, Oslo and Stockholm

Nord Pool, 2005, "Negative Prices in the Elspot market", Nord Pool message, www.nordpool.com

NVE (Norwegian Water Resource and Energy Directorate), 2005, "Leverandørskifteundersøkelsen 4. kvartal 2004" (Report on retail switching 4th quarter 2004), www.nve.no

OECD, 2005, "The Benefits of Liberalising Product Markets and Reducing Barriers to International Trade and Investment: the Case of the United States and the European Union", OECD Economics Department Working Paper 432, Paris

OFGEM, 2004, "2003/2004 Electricity Distribution Quality of Service Report", OFGEM, London

Oren, Shmuel, 2005, "What is a natural capacity mechanism that will meet generation adequacy need s with minimal interference in energy markets?", conference paper, EPRI conference on Global Electricity Industry Restructuring

- in Search of Alternative Pathways, May 2005, http://www.lrca.com/events/EPRI-AC/

Pfeifenberger, Johannes, Joseph Wharton & Adam Schumacher, 2004, "Keeping up with Retail Access? Developments in U.S. Restructuring and Resource Procurement for Regulated Retail Service", *The Electricity Journal*, vol. 10, pp. 50-64

PJM Interconnection, 2005a, "PM Annual Report 2004", www.pjm.com

PJM Interconnection, 2005b, "2004 State of the Market", www.pjm.com

PVO, 2005, "Annual Report 2004", http://www.pohjolanvoima.fi

Rassenti, S. J., V. L. Smith & B. J. Wilson, (2001), "Turning off the Lights", *Regulation*, v. 24, no. 3, Fall 2001, The Cato Institute

South Australia Government Electricity Taskforce, 2001, http://www.treasury.sa.gov.au

Speckler, Jörg, 2005, "The Feasibility of Independent Power Production in Germany", conference paper, PowerGen Europe 2005 - Market Liberalisation: Lessons, realities & opportunities, Penn Well

Sydvest Energi, 2005, "Årsrapport 2004" (Annual report 2004), www.sydvestenergi.dk

Svenska Kraftnät, 2002, "Effektförsörjningen på den öppna elmarknaden" ("The capacity supply in the open electricity market"), www.svk.se

de Tomás, José Antonio, 2005, "Security of Energy Supply for Specific Technologies", presentation at IEA/NEA workshop *Security of Energy Supply for Power Generation - May 2005*, www.iea.org

TVO, 2005, "Annual Report 2004", http://www.tvo.fi

Vilnes, Olav, 2005, "Chasing the Insiders", *Montel Powernews*, No 3, Oslo.

JOUVE
11, Bd de Sébastopol - 75001 PARIS - Imprimé sur presse rotative numérique
N°398217J - Dépôt légal : Avril 2006 - Imprimé en France